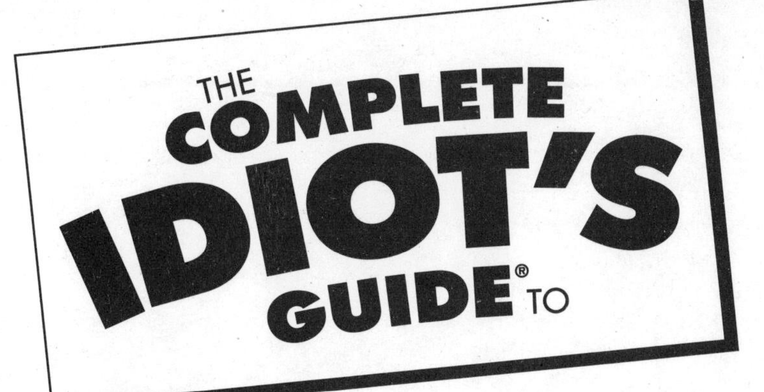

Understanding Cloning

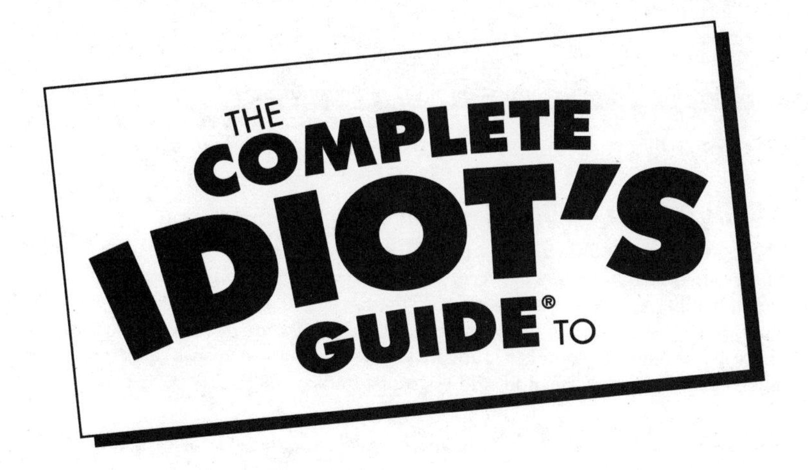

Understanding Cloning

by Jay D. Gralla, Ph.D., and Preston Gralla

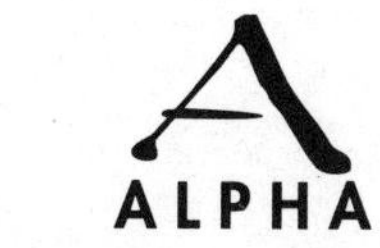

A member of Penguin Group (USA) Inc.

We are who we are because of our parents—no need to debate nature versus nurture, because they provide us with both. So this book is for them, and for Eliot as well, who if he were alive today would have had his name on the cover alongside both of ours.

International Standard Book Number: 1-59257-148-4
Library of Congress Catalog Card Number: 2003116924

06 05 04 8 7 6 5 4 3 2 1

Interpretation of the printing code: The rightmost number of the first series of numbers is the year of the book's printing; the rightmost number of the second series of numbers is the number of the book's printing. For example, a printing code of 04-1 shows that the first printing occurred in 2004.

Printed in the United States of America

Publisher: *Marie Butler-Knight*
Product Manager: *Phil Kitchel*
Senior Managing Editor: *Jennifer Chisholm*
Acquisitions Editor: *Mikal Belicove*
Development Editor: *Michael Koch*
Senior Production Editor: *Billy Fields*
Copy Editor: *Sara Fink*
Illustrator: *Chris Eliopoulos*
Cover/Book Designer: *Trina Wurst*
Indexer: *Brad Herriman*
Layout/Proofreading: *Mary Hunt, Ayanna Lacey*

Contents at a Glance

Contents

Appendixes

Foreword

Imagine a world where cells can be created for therapeutic treatment of certain diseases or whole organs can be generated for transplants. Extinct species of animals could be restored to earth's population. Humans could duplicate themselves or make exact copies of lost loved ones. While these things may seem a bit far-fetched, cloning could one day make them a likely possibility.

Developments in cloning and bioengineering techniques have changed life as we know it. With advances in bioengineering we can have everything from designer drugs to designer food. Researchers have been able to create synthetic skin, plants that produce human blood proteins, and bacteria that protect against cavities. Since the cloning of the sheep Dolly, techniques have been developed for cloning mice, goats, cattle, and other mammals. These successes have opened the door to the possibility of one day cloning humans.

All of these accomplishments are remarkable, however a question that is important to consider is, should we clone?

This question is most commonly asked in reference to human related cloning, specifically embryonic stem cell research. Should we initiate life through fertilization, only to stop the process long enough to harvest the needed materials? Are we sacrificing the lives of some for the lives of others, and should we have that right? For many, ethical and religious beliefs make it impossible to accept these types of cloning methods.

Since cloning is fraught with such ethical and moral dilemmas, one must ask, why is cloning so important? In my opinion, it is the search for eternal life and youth that leads us down this path called cloning. We all want to look younger, feel younger, and live longer. We want to be free from illness and disease. Many see cloning as the "fountain of youth." It has become the key to unlocking the mysteries of aging, not just prolonging life, but also improving the quality of life. It's a well-known fact of life that all living things must die. However, many hope that cloning will become the tool that removes death from the equation of life. To some, the question is not why clone, but why not? Cloning gives us the hope that we can one day make immortality a possibility. To those, the benefits of cloning far outweigh the risks.

In *The Complete Idiot's Guide to Understanding Cloning*, Jay and Preston Gralla delve into everything from the ethical issues involved in cloning, to environmental fears concerning genetically engineered foods, to the enormous political impact that the issue of cloning has had on the world. They provide solid information about DNA, genes, cells, cloning, bioengineering, and much more, in a noncomplicated manner.

If you are looking for an overly technical explanation of cloning, you won't find it here. What you will find is a well written, easy to understand book on cloning and related topics. Jay and Preston do an excellent job of not only explaining the what of cloning, but the why as well.

—Regina Bailey
About.com Guide to Biology (biology.about.com/)

Regina Bailey is a science educator who specializes in making science accessible and understandable. Regina holds a Bachelor's degree in biology from Emory University and works for a web content development firm that specializes in providing biology and science content. She is also active in the production of science content for home-schoolers.

Introduction

There is no doubt what technologies will define our time for future generations: cloning and biotechnology. The advances made in the past decade have been tremendous and they are only a taste of what is to come. Virtually every aspect of our lives—from what we eat, to our medical care, to how we live and work, will one way or another be affected by them.

The technologies will most likely have an even bigger impact on how we think about ourselves and our place in the universe. True, they will lead to better medical care, cures for currently incurable diseases, and most likely a cheaper and possibly more nutritious food supply. But probing the mysteries of life, of personality, of what makes each of us unique, will lead to internal questioning as well. If a person can be cloned from our genes, for example, then what makes each of us a unique individual? If we humans can literally create new life forms with the same ease with which we mix and match DNA, how are we different from God? In fact, if humans can finally understand the mysteries of what creates life, what place does God have in our lives?

For these and other reasons, cloning and bioengineering have become perhaps the most debated issues of our times. They are probably the most misunderstood technologies as well.

In this book we'll show you how cloning and associated technologies work and we'll also explain how genetics, DNA, and the cell itself works. Based on that, we'll delve into the thorny issues related to those technologies and explain the points of view of people on all sides. By the end, you'll fully understand the science, as well as know what people on all sides of the debate have to say. Then you'll be able to make up your own mind. And considering how many cloning and bioengineering claims are made these days, you'll also be able to separate the facts from the hype.

How to Use This Book

To help you understand cloning and biotechnology we've divided this book into four parts. We start off by teaching you the basics of DNA, cloning, and genetics, and then move on to showing you how cloning is actually done. From there we cover the controversies around cloning, and then look at bioengineering, genetics, and the future.

Part 1, "Cloning, DNA, and Genetics Basics," covers the very basics—teaching you about heredity, and how chromosomes and genes work. We show you how genes work with the machinery of the body to build plants and animals and maintain them, and explain DNA's role in it all. You'll learn about genetic diseases as well, and take a look inside DNA's famous "double helix" molecule.

Part 2, "How Cloning Works," gives you the lowdown on how plants and animals are cloned and the problems associated with cloning. We show you how Dolly the Sheep was cloned and how other plants and animals are cloned. We discuss therapeutic stem cell cloning, in which no real "clone" is created, but in which embryonic tissue is used to potentially cure many currently fatal diseases. And we look into how humans might be cloned—and see why some scientists say that this may never be possible.

Part 3, "The Controversies Around Applying Cloning," looks at the wide variety of issues surrounding the cloning debate. You'll read the cases for and against human cloning and therapeutic cloning and find out what political and religious leaders have to say about it. We'll explore related matters such as whether you can patent life and whether animals should be cloned so that they can provide organs for human transplantation.

Part 4, "Bioengineering, Genetics, and the Future," examines the many controversies swirling around biotechnology, from genetically modified foods to the use of gene therapy, and the role played by giant "pharming" companies such as biotech and pharmaceutical companies. This is a debate that predates the debate about cloning, and will most likely go on well past when the cloning issue is ultimately resolved.

The book also includes a glossary that defines terms related to cloning, DNA, genetics and biotechnology, and an appendix that lists the best websites and books where you can go for more information.

Extras

To help you understand cloning, biotechnology, and DNA, this book also gives you extra bits of information and resources. They are sprinkled throughout the book in these boxes:

BioDefinition

These boxes define a term related to DNA, cloning, and biotechnology.

BioWarning

CAUTION

These boxes alert you to a danger related to biotechnology, cloning, or genetics.

BioSource

These boxes point you to places you can go to find more information about a topic, frequently on the Internet, but also point to other books and other places to get information as well.

BioFact

These boxes point out extra noteworthy facts about DNA, cloning, and biotechnology. Sometimes they're little known facts, sometimes interesting tidbits of knowledge, and sometimes revealing stories.

Acknowledgments

As with all books, this was a group effort. Thanks to acquisition editor Mikal Belicove for trusting us with the project, development editor Michael Koch for his sharp eyes and even sharper virtual pencil, copy editor Sara Fink for making sure the writing was as clear as possible and to remind us of all we forgot about grammar, and production editor Billy Fields who shepherded the project through its long and winding road to publication.

Trademarks

Part 1

Cloning, DNA, and Genetics Basics

Sometimes it seems that you can't read the day's headlines unless you've first earned a Ph.D. in biochemistry. Bioengineered foods, cloning, DNA fingerprinting … the list of subjects pertaining to biology and chemistry goes on and on.

In fact, there's no need to get your Ph.D. if you want to understand the day's news—just read this section. It covers all the basics you'll need. You'll learn how genetics works, what a DNA molecule is and how it builds bodies, how cells work, and more. So save yourself five years and several thousand dollars and read this section instead of going back to graduate school.

Chapter 1

Bring on the Clones

In This Chapter

- An introduction to DNA
- How cells work
- An introduction to cloning
- Some words on bioengineering
- Pros and cons of cloning and bioengineering

If centuries can be known by the kind of science and technology that most affected the lives of people, then the twenty-first century may well be ultimately known as the century of cloning and bioengineering. The technologies have already changed what we eat, the future of medicine, global agriculture, policing and crime, the concept of what constitutes weapons, and even far-flung sciences such as anthropology, and much more. There isn't a facet of our lives that these technologies haven't touched.

Possibly more important, cloning touches on religion, ethics, and morality, and goes to the very heart of what it means to be a unique human being. If that isn't enough, it also brought us the Raelians, a pro-cloning UFO cult run by a balding, middle-aged, pony-tailed former racing car journalist from France who believes that human beings were first created in test tubes on a faraway planet, and then brought to Earth in UFOs. Next to that theory, $E=MC^2$ pales in comparison.

In this chapter, we'll take a quick look at DNA, cloning and bioengineering, and see why they're the most important technologies of the century.

DNA: The Mother of Us All

The twenty-first century may be known as the century of cloning and bioengineering, but that technology has its basis in a startling breakthrough made in the mid-twentieth century in 1953 by James Watson and Francis Crick—a description of the structure of the DNA molecule. Watson and Crick didn't actually discover DNA; its general role in genetics was understood before their breakthrough. But uncovering its molecular structure was the first step in understanding how it works, and understanding how it works in turn led to all the advances in genetics and bioengineering that have taken place since. Not to mention the Raelians.

BioFact

There's a bit of controversy surrounding whether a third person should be given credit for discovering the structure of DNA. Rosalind Franklin did initial work to understand the structure of the molecule, and an X-ray picture that she took of the molecule, called Photograph 51, was shown to Watson without her knowledge, probably innocently by Franklin's colleague, Maurice Wilkins. Watson says that the photograph went a long way toward giving him clues to unwrapping the mystery of the molecule. In fact, he writes in his book *The Double Helix*, "The instant I saw the picture my mouth fell open and my pulse began to race." Franklin died before the Nobel prize for DNA structure was awarded.

A DNA Snapshot

What Watson and Crick found was one of the more amazing molecules in creation. Shaped like the now-familiar double helix as shown in the following figure, a single strip of DNA, if uncoiled, would be six feet in length. Yet when folded in its natural shape it can fit into a size measured in the trillionths of inches so that the molecules can fit into the nuclei of every one of our cells.

The "working parts" of the molecule can be described with four chemical letters—A, T, C and G—and yet, the endless variations of those letters along the molecule's length give rise to the endless variation of life on our planet. Encoded in those letters are what makes a worm a worm, a mouse a mouse, a man a man; what gives one of the authors of this book brown hair (Preston) and the other author blonde hair (Jay);

what makes some people short and some tall; what makes each and every one of us a unique individual. In short, if there is a Book of Life, it is written with the letters A, T, C and G and is contained along the length of the DNA molecule.

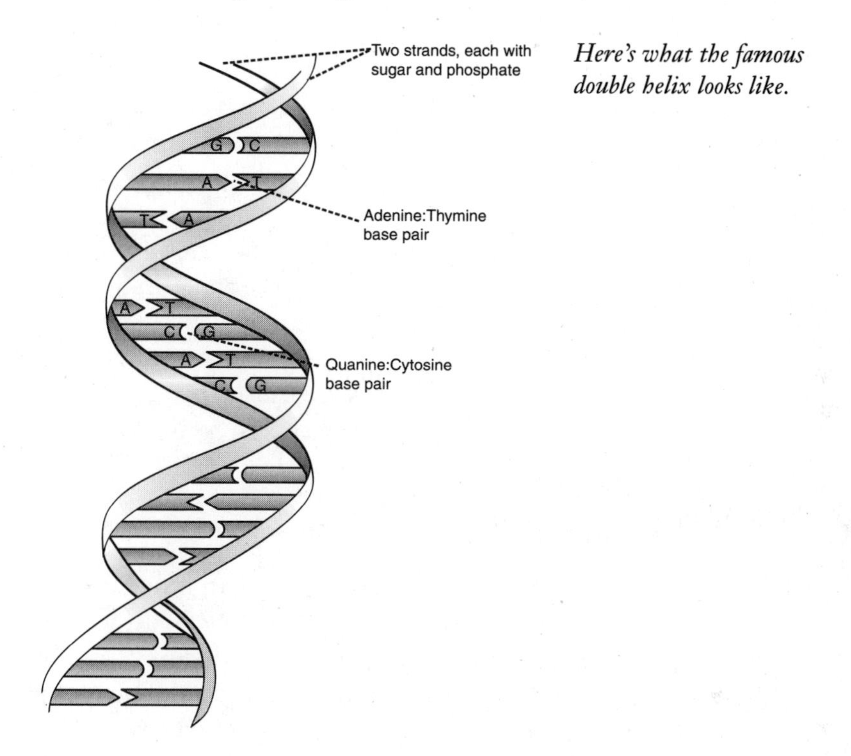

Here's what the famous double helix looks like.

What DNA Does

Along each DNA strand are found genes, which carry the genetic information required to build each of us. This genetic heritage is passed from generation to generation. Some of what is passed to each generation by genes is visible—for example, the color of someone's skin, eyes, and hair. But the vast majority of the work of genes is invisible, and you take it for granted. For genes don't only give instructions for hair color; they literally tell your body how to build itself and how to function.

The total number of genes in the human body is somewhere in the 30,000 range. But before you start puffing out your chest and strutting around with pride because of that vast number, consider this: We have only about three times as many genes as a fruit fly. And we don't even have wings! (On the other hand, fruit flies don't have toenails, so I guess we're even.)

Much of life shares a similar genetic heritage—our genetic makeup, for example, is surprisingly similar to that of mice. In fact, if you could take pieces of human DNA, cut them up and discard much of them, you'd be able to assemble a reasonable facsimile of the DNA makeup of a mouse. So the next time someone asks you whether you're a mouse or a man, you may have some thinking to do before you answer. Of course, the actual organisms of men and mice are dramatically different from one another. Still, our DNA isn't as different as you might expect.

DNA is the foundation of all life, yet at the core level, what genes do is amazingly simple, even mundane. The sole purpose of each gene is to create a specific protein or family of proteins, often *enzymes*. That's it. Nothing more and nothing less. These proteins do the heavy lifting of building bodies and making them function. So if a gene mutates, it might not be able to create the specific protein that it was supposed to create, and the consequences of that one changed or missing protein can cause deadly diseases. Most genetic diseases are caused by making the protein but in a slightly altered form that does not function properly.

BioDefinition

An **enzyme** is a special kind of protein that causes or speeds up vital chemical reactions in cells and the human body.

Where Cells Come In

DNA, of course, can't do its work alone. For that, it needs specialized workplaces, and those are the cells in your body. Every cell in your body, whether it's a heart cell, a muscle cell, a liver cell, a toe cell, or any other kind of cell, contains the exact same DNA in its nucleus. (Actually, there's a slight exception here: Sperm and egg cells only contain half of your DNA, and we'll explain why that is in Chapter 2.)

In the middle of each cell is an object called the nucleus, which is command-central and is in charge of all the cell's functions. Inside the nucleus are chromosomes, which are long strands made up of DNA. Along the DNA are individual genes. The DNA issues instructions to the cell to create specific proteins. One of the greatest mysteries in life is why only certain genes are active in certain cells. For example, your stomach cells don't create proteins used only in heart cells (although if they did, it would give new meaning to the saying, "my heart was in my stomach"). Instead, they create only proteins necessary for the functioning of your stomach. How specific genes are turned on and off throughout your body remains a mystery to a great extent, although scientists are learning more about it every day.

Whenever a cell replicates, the chromosomes in its nucleus copy themselves, and each new cell gets a complete copy of all the original cell's DNA. In this way, every cell in your body (again, with the exception of sperm cells and egg cells) has your complete genetic code. Amazingly, the new cells know exactly which genes to turn on and keep off.

Here Come the Clones

We are all born with our unique genetic code, our own specific mix of genes, that makes us uniquely us, unless we are one of a pair of identical twins, or triplets, or some other-lets. (Each of them have the same genetic makeup.)

Identical twins occur when a single embryo splits into two identical embryos. Each of those embryos have the same genetic makeup as the original. They came from the same egg that was fertilized by the same sperm, and so have the same genes.

If nature can create twins, scientists have wondered, why can't man do the same and create *clones*? In fact, a rudimentary method for creating animal clones mirrors the way that nature creates identical twins. In the procedure, scientists divide an animal embryo in two during the early stage of its development, and in some cases, each embryo then develops into adult animals. Instant clone!

BioDefinition

The word **clone** comes from the Greek word *klon*, which means twig or shoot. That makes sense, since one can use twigs or shoots to make copies of plants.

Things in the cloning world have gotten much more sophisticated than that, though. In 1997 the biggest cloning breakthrough took place when scientists at the Roslin Institute near Edinburgh, Scotland cloned a sheep using a method called *nuclear transfer*. First, an unfertilized egg was taken from a ewe, and its nucleus was removed. Because the nucleus of a cell contains its DNA, that means when the nucleus is removed, so is the egg's DNA. The ewe was a Scottish Blackface sheep, which, as its name implies, has a black face.

Next, scientists took mammary cells from another sheep—another female but a Finn Dorset sheep, which has a white face. They planned to clone this sheep by moving its DNA into the DNA-less egg. They treated the mammary cells, then placed one cell next to the egg missing a nucleus from the Blackface sheep. Pulses of electricity were sent through the two cells, fusing them into a single cell. (Shades of Frankenstein!) The egg cell now had the nucleus of the mammary cell—and more importantly, it had the DNA from the mammary cell. More electricity was sent through the fused, single cell, and the egg cell began dividing just as it does normally in nature. (The second jolt of electricity mimics the natural jolt of energy that occurs when an egg is fertilized, and which helps to start the egg dividing.)

Scientists let the embryo develop for about a week, then implanted it into the uterus of Blackface sheep. The embryo developed normally, and ultimately a sheep was born—an exact clone of the Fin Dorset sheep, with a white face. Dolly entered the world, and the world hasn't been the same since. Dolly has since died, but her fame lives on.

BioFact

You might have heard the term the *human genome* and wondered what it means. A genome is the sum total of all the genes in a given species—and the genome map identifies them all and shows exactly where each gene is located on chromosomes. The Human Genome Project was a program that mapped the human genome. Using this map, researchers hope to be able to get a firmer understanding of every gene's function, and armed with that information, may be able to cure many genetically linked diseases.

A variety of other cloning techniques have been developed since then. For example, mice have been cloned by directly injecting a nucleus from a cell from one mouse into an egg cell from another mouse whose nucleus has been removed.

What Is Bioengineering?

The other big deal in the DNA world is bioengineering, which to date has had a much bigger impact on the way we live than has cloning. (Although to my knowledge, no bioengineering UFO cults have yet sprung up.)

In bioengineering, scientists essentially mix and match genes from one species with another, or in some other way manipulate living genetic material. It's affected you and probably everyone you know. Whenever you eat food with corn or soybeans in it, you've probably eaten foods that have been bioengineered. Corn and soybean plants often contain bacterial genes that make their growth more resistant to insects and herbicides. Additionally, a gene in Flavr Savr tomatoes—tomatoes which are commonly sold in many grocery stores and supermarkets—has been altered so that the tomato ripens more slowly, allowing it to be more easily shipped long distances without becoming overripe. (Of course, there are those who say that a price was paid for the engineering—making some tomatoes taste like cardboard.)

Bioengineering has been used to create drugs as well. For example, genes that produce particular human proteins have been implanted into bacteria and animal cells, and those cells then produce the proteins in very large amounts. These proteins are used in drugs including insulin and human growth hormone. Without bioengineering, the drugs would not be able to be made in such quantities and would be much more scarce and expensive.

In the future, scientists believe that using bioengineering, we will be able to create "designer drugs" that are specially engineered for someone's particular genetic makeup or disease.

What Are the Issues with Cloning and Bioengineering?

Very few people dispute that cloning and bioengineering are remarkable technologies. And most people also agree that advances in DNA research have some very real benefits as well.

But there the agreements end. The technologies have the potential to drastically change the world around us, the way we live and work—and perhaps most important, raise potentially troubling questions about identity and what it means to be a human being. Not to mention the Raelians.

So in this section, we'll take a quick look at some of the issues surrounding the technologies. We'll look at all the issues in greater detail in the rest of the book, but we'll start here with a quick overview.

Pros and Cons of Bioengineering

The debate about bioengineering has been raging for several decades, but it picked up steam in 1991 when Calgene Inc., of Davis, California, asked the Food and Drug Administration to approve its Flavr Savr tomato, the first bioengineered agricultural product.

Proponents of bioengineering say that the technology may solve age-old problems, including disease and hunger. Specifically, here are some of the benefits they claim the technology can offer:

- **It can help stop world hunger.** Bioengineered plants can resist disease and pests, lengthen the growing season of plants, increase crop yields, and can put vital nutrients into foods that are otherwise expensive to provide—for example, bioengineering rice so that it is high in protein or African yams so that the majority are no longer destroyed by blight before they can be eaten. Similarly, bioengineered animals can be raised less expensively and more easily than those created by nature.

BioFact

The next time you see the word *pharming* in use, don't assume someone doesn't know how to spell. It's a pun, and an increasingly common term. It's a combination of farming and pharmaceuticals. It refers to the use of transgenic animals, plants, or cells to produce pharmaceutical products.

- **It can produce inexpensive drugs.** Drugs can be very expensive to produce, and in some instances, there may be no way to produce large quantities of a given pharmaceutical. Bioengineering can produce large quantities of important drugs at lower costs than traditional methods.
- **It can help find cures for diseases.** Scientists everywhere use bioengineering techniques in medical research, and may even be able to use bioengineering to cure diseases, for example in the use of gene therapy in which genes are introduced into humans to help cure disease.
- **It can solve the problem of organ transplants.** Thousands of people are on waiting lists for organ transplants; people die because there aren't enough transplantable organs available. Some scientists believe that we may ultimately be able to raise bioengineered animals that will grow organs that can be safely transplanted into humans.
- **There are many benefits we can't even dream about.** Any powerful technology, once introduced, ends up being used in ways that its inventors never imagined. Proponents of biotechnology say that the same will happen with bioengineering.

On the other side of the issue are those who say that bioengineering is inherently a dangerous technology that will ultimately do far more harm than good. Here are some of their main arguments:

BioFact

The fields of DNA research, cloning, and bioengineering have given rise to a series of punning terms. One of the longest lasting one is "Frankenfoods"—a term that detractors give to bioengineered foods, and that implies that they are dangerous and beyond the control of their creators.

- **Biofoods are potentially unsafe.** They may introduce proteins or other chemicals into foods that can cause severe allergic reactions in people who don't realize the foods have been bioengineered and contain the proteins. For example, a bioengineered soybean had a gene inserted into it from Brazil nuts—and that gene created a protein associated with Brazil nuts. People allergic to Brazil nuts never would have imagined that they could also be allergic to soybeans, and so the soybean could have been dangerous to them. Fortunately, the soybeans were never grown commercially.
- **Bioengineering concentrates power in large corporations.** Large corporations can easily afford to carry out bioengineering on a large scale, and so the technology concentrates economic power in them, say detractors. These large

companies often introduce bioengineered crops that are sterile so that the crop's seeds can't be used to plant the next year's crop. Critics contend that this harms economically stressed farmers, particularly in poor Third World nations.

- **Bioengineered plants and animals can get loose in the wild.** What happens if pollen from bioengineered plants spreads to non-bioengineered plants? Will it make them sterile? Will it kill out non-bioengineered plants? No one is really sure, say critics, and the potential danger outweighs bioengineering benefits.
- **Animals shouldn't be abused.** Animal-rights proponents say that animals shouldn't be treated as organ factories for humans, and so bioengineering technologies pointed at that should be banned.
- **There are many dangers we can't even dream about.** Critics offer the mirror-image of proponents here. They say that we can't ever know the dangers of a technology until well after it's been released, and so there are potential dangers we don't know about.

CAUTION

BioWarning

One of the most controversial uses of biofoods is the use of Bovine Growth Hormone (BGH), which is injected into cows to force them to produce more milk. Cows naturally produce a certain amount of BGH, but large amounts are produced when the gene for creating the hormone is inserted into bacteria, which then become hormone-creating factories. The resulting hormone can then be injected into cows to increase their production of milk. Some worry that the hormone may be unsafe for human consumption, and the technology has been banned in some countries. Many dairies refuse to use the hormone because they worry about a consumer backlash against it.

Pros and Cons of Cloning

Cloning of organisms is an even more controversial technology than bioengineering. To date, it has not been used commercially as much as bioengineering, but it has engendered an enormous amount of controversy. Very few people recommend cloning entire human beings and even scientists are mostly united on this issue. However, many scientists favor using discarded embryonic cells for medical research—so called *therapeutic cloning*. In this procedure the embryonic cells are collected and allowed to reproduce in laboratory dishes. Those cells can then be used for research purposes and to fight disease.

Here are some of the benefits proponents claim that therapeutic cloning can offer:

- **It may provide a cure for Parkinson's disease.** It's possible that the cloned cells could be grown into nerve cells and so cure those who suffer from this debilitating disease.
- **It may help cure diabetes.** The cloned cells could be turned into insulin-secreting cells, and so cure diabetes.
- **It could help cure burn patients.** The cloned cells could be grown into skin that could be used for burn patients.
- **It may help clone organs.** Countless people die each year because they need organ transplants, but no organs are available. It may be possible to clone entire organs using therapeutic stem cells, although this use would be far in the future.
- **It can help with many other diseases.** There are countless diseases that could theoretically benefit directly from the use of cloned therapeutic stem cells.
- **It will help in ways that we can't imagine.** Learn how cells develop from embryonic cells into the myriad cells of our body, and to a great extent, you'll unlock one of the great mysteries of biology—and life. That could lead to countless medical breakthroughs.

There are even people who go beyond that, and say that cloning entire human beings would serve a purpose. If a child died, they say, for example, parents could clone that child, which would be in essence a twin of the dead child.

There are those who believe that any form of cloning, including therapeutic stem cell cloning, should be banned. They warn of the following:

- **Cloning an entire human being is dangerous.** As you'll learn in Chapter 10, cloning carries with it many medical dangers, and animals that are cloned frequently die at a young age with a variety of medical problems. To do that to a human being would be immensely cruel.
- **Those who would clone a child are mentally unprepared to be parents.** Anyone who would want to clone a child is clearly not a capable parent, because of the dangers they would put the child through, this thinking goes. And those who would want to clone themselves are even more unbalanced and unsuitable parents.
- **Cloning human beings is immoral.** It breaks a wide variety of ethical, religious, and moral taboos.

- **Cloning therapeutic stem cells violates religious or moral teachings.** There are those who say that religious beliefs ban even cloning therapeutic stem cells. Some who worry that embryos would be created specifically for harvesting find this morally repugnant.

We'll look at all these issues, and the issues surrounding bioengineering in much more detail in Part 4.

The Least You Need to Know

- DNA contains the genetic material of all plants and animals. All of an organism's DNA is called its genome.
- The nuclei of cells contain chromosomes, which is made up of DNA. Genes are located on DNA molecules.
- Each gene carries the information necessary for creating a single family of proteins or enzymes.
- The genome map shows all the genes in a given species.
- Bioengineering can mix and match genes from different species, or manipulate the genes of plants in animals in some way.
- When you clone a plant or an animal, you make a copy of all its DNA and create a plant or animal with that identical genetic makeup.

Chapter 2

Understanding Heredity, Chromosomes, and Genes

In This Chapter

- All about Mendel, the father of genetics
- How Mendel developed his genetic theories
- How heredity works
- What genes and chromosomes are
- How genes pass from generation to generation

Why do some people have blue eyes and some brown eyes? If a pea plant has one tall parent and one short parent, will it grow up to be tall? What's a gene? And how does heredity work, anyway?

If you want to understand how cloning works, you first need to understand heredity, genes, and chromosomes. That's what we'll cover in this chapter. To explain it all though, we need to hop into the wayback machine, and head back to the mid-nineteenth century, at the time when modern genetics began.

In the Beginning Was Mendel

By the middle of the nineteenth century, some people thought they had heredity all figured out. You were a mix of the features of your mother and father, so if you had a tall father and a short mother, you'd end up as medium height. Take the average of what your parents were, and that's what you'd be.

Of course, there were quite a few inconvenient facts the theory couldn't account for, such as eye color—if one of your parents had blue eyes and the other brown eyes, why did some children have blue eyes and some have brown, and none have a mix of the two? And what color *do* you get if you mix brown and blue, anyway? Clearly, some work needed to be done on the theory.

You might expect that the leading scientists of the day would solve the conundrum of heredity. But as is often the case in science, the answer came from an unexpected quarter—from Gregor Mendel—an Augustinian monk in Austria fiddling around with peas in his garden, who couldn't even pass the exam required to teach high school science.

But Mendel may have had an advantage in his upbringing that other scientists did not: He grew up on a farm, and farmers then and now are above all practical people, guided by experience and observation rather than by theory. (After all, you can theorize all you want about the best way to grow corn, but words won't do the trick—to do it, you need rain, soil, and nutrients, not syllables.)

For millennia, farmers had been breeding hybrids, crossing two varieties of animal or plant *species* in order to get more efficient, more useful, or "better" animals or plants. Breeding was largely a matter of trial and error—farmers couldn't always predict what would happen when varieties were bred. Sometimes the offspring was like one of the parents, some time like another of the parents, and other times a blend of the two. And on rare occasions, parents produce an offspring with a characteristic completely unrelated to either parent.

BioDefinition

A **species** is a group of animals or plants that share the same, common characteristics and are capable of interbreeding and producing fertile offspring. So, for example, a horse and a donkey can produce a mule as offspring, but a mule is sterile and can't reproduce—and that's one reason why horses and donkeys are different species. Things are not always black-and-white when it comes to determining whether plants or animals are of the same species though, and not uncommonly organisms that were once thought to be of the same species turn out to be of different species. And sometimes plants or animals that were thought to be of different species turn out to belong to the same one.

When Mendel took over responsibility for caring for his monastery's garden, he set out to learn about why offspring inherited some characteristics of their parents, but not others, and how those traits were passed along.

The Secret Is in the Peas

In his quest for the perfect plant to study all this, Mendel turned to the lowly pea. It wasn't that he craved split pea soup, or that he was looking for peas on earth. (Sorry for the pun; we couldn't resist.) It was that the pea was perfectly suited for research into heredity. There were a number of varieties of peas that had very noticeable characteristics, and so the plant was perfect for cross-breeding—it would be very easy to see the results. For example, there were tall and short varieties; green and yellow varieties; varieties that produced smooth peas and varieties that produced wrinkled peas; there were varieties with white flowers and varieties with purple flowers … and that's just for a start. Because of the obvious differences, it would be easy to track the results of what happened when he crossed different varieties—all he had to do was look at the offspring.

In addition to that, peas are simple plants, and are easy to cross-breed using pollen from one plant, and an "egg" from the other. They're also quick growers, and so he didn't have to wait a long time after cross-breeding his plants to see the results.

Mendel was a meticulous kind of guy (which is pretty much written into the job description for monks) and so before he began his experiments, he wanted to make sure that he started off with plants that had stable sets of characteristics. So he made sure that when the tall varieties fertilized themselves they always produced tall plants, when the short varieties fertilized themselves they always produced short varieties, and so on.

Mendel's First Results: Let's Get Tall

As part of his first set of experiments, Mendel crossed a tall plant with a short plant. According to the theories of the time, the offspring should have all been medium-sized plants.

But something very curious happened. The offspring weren't medium-sized plants. In fact, every one of them, without exception, was tall. It didn't matter which plant contributed the pollen and which the egg—every time a tall plant was crossed with a short plant, the result was a tall plant.

BioFact

Mendel was quite clever in the way that he fertilized his pea plants. Normally, peas are self-fertilizing. An organ called *anthers* in the pea flower produces pollen, and another nearby organ in the flower called *stigma* produces eggs. Pollen from the anthers falls on the stigma, and so the pea fertilizes itself. In Mendel's experiments, he cut off the anthers of some pea plants while the plants were still immature and had yet to produce pollen. Then, when the stigma had produced eggs, he dusted the stigma with pollen from a different pea plant, one that he wanted to be the "father." Finally, he tied bags around the flowers to keep out any other pollen, and so was able to cross two different pea plants.

Mendel tried the same thing with green pods and yellow pods. And a similar curious thing happened. Rather than producing plants with yellow-green pods, every plant from the union had a green pod.

After other experiments produced similar results, he came up with a conclusion that seems obvious to us, but was revolutionary at the time—some traits are dominant over others, while other traits, which seem to disappear, are recessive. This simple observation is the foundation upon which genetics is based. Not bad work for a monk and a bunch of peas.

On to the Next Generation

One revolutionary discovery in a lifetime wasn't enough for Mendel, so he didn't stop there. He took that first generation of offspring and bred them with each another. So, for example, he took all the tall plants that were the result of crossing a tall plant and a long plant, and bred those offspring with each other.

BioFact

Mendel's findings about genetics are based on his work with peas. But in fact, his first work wasn't with peas, but instead with mice. He wasn't able to make those experiments work, so he turned from the animal kingdom to the vegetable kingdom. The results tasted much better—and he didn't have to clean up after peas, either.

The results were even stranger than they were the first time around. You would expect that all of the resulting plants would be tall—after all, tallness was dominant.

Instead, one quarter of the plants were short, and three quarters were tall. Again, it didn't matter which plant provided the pollen and which the egg—one quarter of the offspring were short and the rest were tall.

He tried the same experiment with peas with other traits, and the results were again the same. In the first generation, one trait was thoroughly dominant and the other apparently disappeared. In the second generation, the recessive trait returned, but to only one quarter of the plants. The dominant trait was found in three-quarters of the plants.

Mendel's Conclusions

So what did it all mean? Mendel concluded that something in the pollen and egg of each plant determines whether it will be tall or short, produce wrinkled peas or smooth peas, and determine all of the other plant characteristics that he studied. He called that special something a *gene*. Furthermore, he said that each plant has two genes for each characteristic, and one of those genes is dominant, and the other recessive. The plant has two genes, because it gets one gene from the pollen and one gene from the egg, he said.

Mendel also concluded that inherited traits are independent of one another. In other words, whether a plant was tall or short had no bearing on whether that plant's seeds were smooth or wrinkled. They each had their own independent genes.

About 150 years after Mendel's discoveries, what he says still forms the basis of genetics. We've since discovered that there are some exceptions to the rules he established—for example, there are instances in which inherited traits are *not* independent of one another—but still, his findings largely hold true.

BioFact

Today, Mendel is hailed as the father of genetics. But in his time, he was little-known, and ignored by the scientific establishment. He published a paper about his findings, but only in a small journal of a local organization called the Brunn Society for the Study of Natural Science. He sent his paper to well-known biologists, but all ignored him. He eventually became abbot of his monastery in 1868 (could it have had something to do with all the pea soup he made?) and all but gave up his scientific experiments. He died in 1884, and had no idea that his work would eventually be hailed worldwide.

Solving the Heredity Puzzle

Now that we know about Mendel and peas, we can apply what he learned to heredity in general. As a review of what we already know:

- Genes determine certain characteristics in peas (and other living things), such as height.

- There are two genes for each of these characteristics. One comes from the father, and one from the mother.
- One gene is dominant and one is recessive. When a dominant gene is present, its characteristic rules.

To help you understand genetics more, here's one more piece of genetic jargon: *allele*. An allele is the form of a gene for a particular characteristic, and a given gene can be one of two alleles. In other words, in peas there are two alleles of the gene for height—one for tallness and one for shortness.

BioFact

Wonder what genes are dominant and which recessive in human beings? Here's a small sample: Brown eyes are dominant; blue eyes are recessive. Color vision is dominant; color blindness is recessive. The capability of curling your tongue is dominant; the inability to curl your tongue is recessive. Why they're dominant or recessive is another matter and one that is only sometimes understood.

About Alleles, Genes, and Heredity

To understand how heredity works, let's step back to Mendel for a minute and take a closer look at why his experiments turned out the way they did. We'll examine his experiments with the heights of pea plants. He started out with two types of plants, tall plants and short plants. He made sure that each of those plants was pure—that is, when the plants self-fertilized, tall plants produced only tall plants and short plants produced only short plants.

The gene for height has two alleles. We'll designate the tallness allele A and the shortness allele a. The following figure shows the original genetic makeup of the tall plants and the short plants.

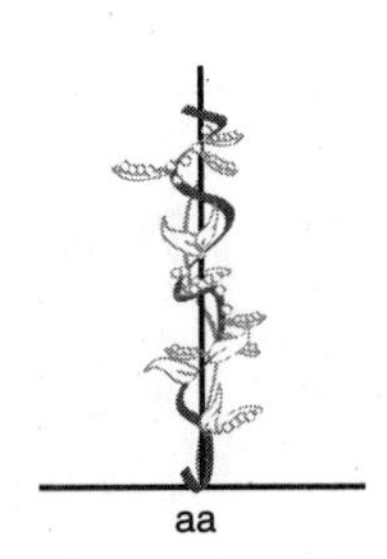

Small or tall: the genetic makeup of Mendel's original tall pea plants and short pea plants.

In his first experiment, Mendel bred a pure tall plant with a pure short plant. In each new plant, a gene from a tall plant was mixed with a gene from a short plant, and so its genetic makeup is Aa, as you can see in the following figure. Because each plant has a dominant A gene, each plant is tall.

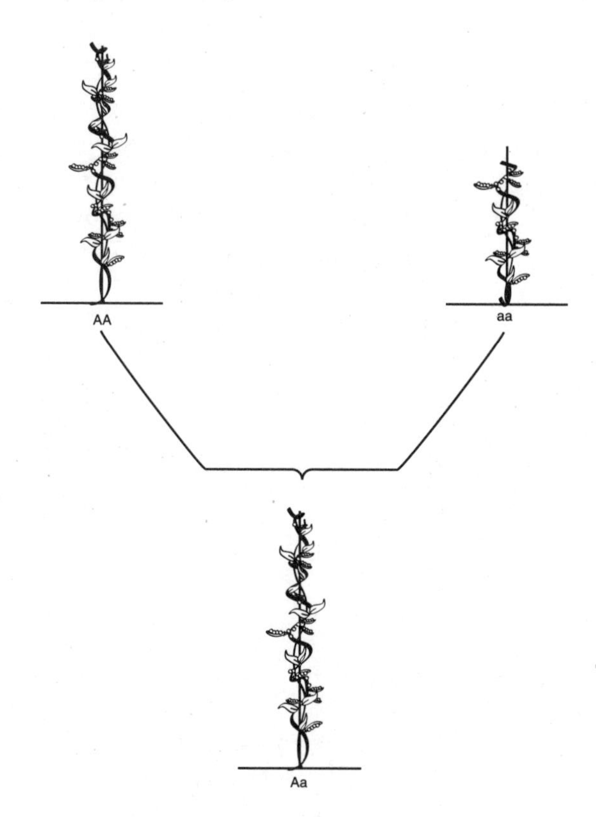

The same but different: The offspring looks like the tall pea plant, but its genetic makeup is different.

All of the offspring have a genetic makeup of Aa. And although those offspring look like their tall parents with an AA makeup they in fact are genetically different—they carry the recessive a gene.

This leads us to a few more pieces of genetic jargon:

- When two organisms have the same genetically determined characteristic, they are the same ***phenotype***. So the AA plants and the Aa plants are of the same phenotype—both are tall.
- When two organisms have the same genes for a given characteristic—when they have the same mix of alleles for it—they are the same *genotype*. So the AA plants and the Aa plants are of different genotypes, even though they look exactly the same.

- When a plant's or animal's two alleles for a given gene are the same, the animal or plant is homozygous for that gene, and if they are different, it is heterozygous for the gene. So AA and aa plants are homozygous, and Aa plants are heterozygous.

Putting it all together means, for example, that the Aa plants are of the same phenotype as AA plants, but a different genotype from them.

Remember that when Mendel bred the second generation of plants, one quarter were short and the rest tall. That's because there are four resulting combinations from mixing an Aa with an Aa: AA, Aa, aA and aa, as shown in the following figure. The dominant A is in three of the four instances, and so three of the four plants are tall.

Here's why the second generation produced three-quarter tall plants and one-quarter short ones.

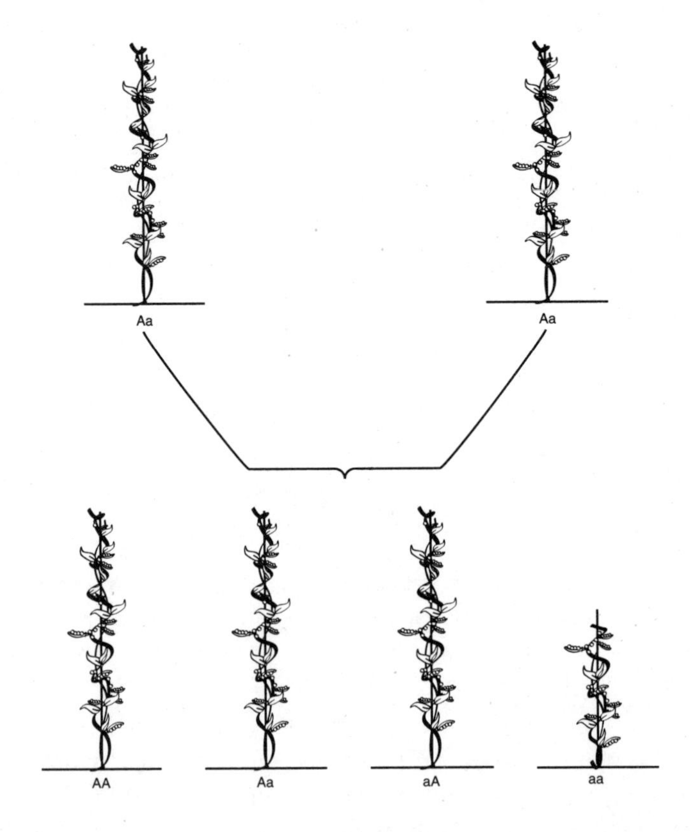

There's a shorthand way for figuring out the results of breeding—draw a square like that pictured in the following figure. Along one side put the alleles for one parent, along another side put the alleles for the other parent, and then just fill in the squares. Instant genetics!

	A	a
A	AA	Aa
a	Aa	aa

Construct a square like this one to figure out the results of cross-breeding.

Mama Don't Take My Chromosomes Away

All this may be well and good, you might be thinking, but after all we're people, not peas so what does this have to do with humans? Well, the same basic rules of genetics apply to people as do to peas; heredity works essentially the same for all plants and animals.

Mendel's theories proved out, but there was one problem with them. What exactly are genes? Mendel's findings were based on observation, but he never actually looked for or found a gene. It took a long time to find them, but eventually we did.

Genes are physical objects, and they're found in all the cells of our body, and in the cells of every living thing. The center of every cell has a nucleus, and coiled inside each nucleus are small objects called chromosomes. Chromosomes come in pairs, so that humans, for example, have 23 pairs of chromosomes, for a total of 46.

BioFact

Each living thing has a certain number of chromosomes—but when a plant or animal has more chromosomes, it doesn't mean that it is more complex than a plant or animal with fewer chromosomes. The actual chromosomes themselves can be different sizes and hold differing amounts of DNA and genes. For example, humans have a total of 46 chromosomes, chimpanzees have 48, dogs have 78, and goldfish have a startling 94. On the other hand, mosquitoes have only 6, while cabbages have 18 and tobacco has 48. The pea plant, the plant that helped Mendel understand genetics, has 14.

Chromosomes are made up of DNA—the basic genetic material of life. (You'll learn a lot more about DNA in Chapter 3 and throughout this book.) DNA molecules are exceedingly long and thin, and along their length are those objects that we've been talking about for this entire chapter—genes.

As you've learned, these genes are responsible for our hereditary characteristics. As you'll learn in Chapter 5, they do this by building molecules called *proteins* and *enzymes*. A single gene builds a single protein or enzyme—and that's all it does.

Amazingly enough, all this protein-building and enzyme-building somehow adds up to a human being (or pea plant, or chimpanzee, whatever the case may be).

Is This Where Sex Comes In?

Mendel bred pea plants by using pollen from one plant to fertilize another plant. But you and I don't come from the pea patch, no matter what your mother and father might have told you when you asked them about the facts of life. So how do genes get passed from generation to generation?

A little while back in this chapter, we explained that all the cells in your body have the full complement of chromosomes in their nuclei—23 pairs of chromosomes for a total of 46 chromosomes. Well, we lied a little. Sperm cells and egg cells are the only two cell types that don't have the full amount. Each has only half the normal number of chromosomes, 23. That's because unlike other cells in your body, they have only one set of chromosomes, not two. There's a special name for sperm and egg cells; they're called germ cells or gametes.

When a sperm fertilizes an egg, their nuclei combine, and so the fertilized egg has the full complement of chromosomes—46—with 23 coming from each parent. The fertilized egg, called a *zygote*, starts developing and splitting. Each new cell has the full complement of 46 chromosomes—that is, until a person reaches maturity and starts producing sperm cells or egg cells, and those two cell types only have 23 chromosomes.

Why a Man Is a Man and a Woman Is a Woman

You've no doubt noticed by now that men are different from women. But what makes them different? No, we're not talking about shopping habits, the distribution of body hair, or the ability to sit in front of a television for 20 hours at a stretch during football season. We mean, why does a man become a man and a woman become a woman?

Not surprisingly, it has to do with chromosomes. Females have a pair of matching chromosomes called X chromosomes. Males, instead, have one X chromosome and one Y chromosome. That one Y chromosome makes all the difference. Most sex-related differences are found in the genes on these chromosomes. So an XX pair creates a female and an XY pair creates a male.

Just to make sure that boy babies and girl babies are produced in the proper proportion, draw a square similar to the one for Mendel's peas and make sure that it all works out. The following figure shows the results—and as you can see, half of the off-spring have XX chromosomes and so are female, and half have XY and so are male.

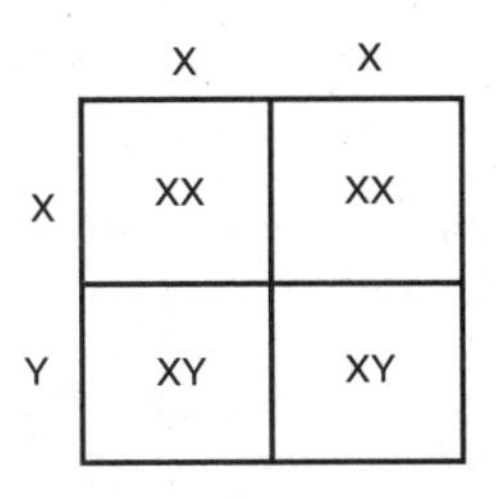

It's a boy! And a girl! And a boy! And a girl!

What Does All This Have to Do with Cloning?

So what does all this have to do with cloning? Everything. If you want to make an exact copy of a plant or an animal, you need to clone all its genetic material—its chromosomes. But you can't clone an egg or a sperm cell, because they only have half the required number of chromosomes, and so you can't build a human being from either of them alone. There are many ways to clone, but commonly, you take the nucleus out of an egg, and in its place put the nucleus of a cell with the full complement of chromosomes from the animal or plant you want to clone. Then you do a little bit of scientific magic, and the egg starts dividing more or less as if it were a fertilized egg. Voilà—you have a clone.

Now, you may be thinking there's an easier way to clone. Why couldn't you make a clone from an egg cell or a sperm cell alone, even though each only has 23 chromosomes? Since only one chromosome in each pair determines a characteristic, why won't that work?

In fact, even though only one chromosome in each pair determines a characteristic, the human body requires the full complement of chromosomes in order to function. Cell division, protein building and the other micromachinery of our body just doesn't know how to deal with the wrong number of chromosomes—you might end up with dividing cells but they could never specify a real organism with different types of cells.

The Least You Need to Know

- Gregor Mendel, the father of genetics, studied the characteristics of peas to better learn how traits are inherited.

- Genes determine characteristics of plants or animals, such as the height of pea plants, or the color of one's eyes.

- For many traits, there are two possible genes—one dominant and one recessive. The dominant gene determines the characteristic of the plant or animal. So a pea plant that has a dominant tall gene and a recessive short gene will produce a tall plant. Only plants with two recessive short genes will produce a short plant.
- Chromosomes contain many genes, which are made up of DNA. Chromosomes are found in every cell of plants and animals, and come in pairs.
- Plants and animals get one set of chromosomes from the father and one set from the mother.

Chapter 3

How Cells Work

In This Chapter

- About cells, tissues, and organs
- A look inside a bacterium
- All the parts that make up a cell
- How cells duplicate themselves
- How sex cells duplicate themselves

If you're going to understand cloning, bioengineering, and DNA, you first have to understand how cells work. DNA by itself is just a molecule, but put it into the center of a cell and the action starts.

In order for the DNA to function, it requires cells to do all the work. So in this chapter, you'll learn how cells work, how they keep you alive, and how DNA fits into the scheme of life.

Cells: Nature's Littlest Factories

Every living thing, from the tiniest bacteria to plants to human beings, is made up of microscopic units called *cells*. Some living things, such as bacteria, are composed of only a single cell, while others, such as human beings, are made up of trillions of them.

From the outside, cells in a living thing may look extraordinarily different from one another because they perform drastically different functions. Nerve cells, for example, are long and thin, while cells in the retina of the eye are shaped like cones or rods. And those are just two examples of many throughout your body, and throughout the animal and plant kingdom.

Cells usually don't do work on their own. Instead, they band together into groups called *tissues*, and tissues band together into organs. So, for example, a diverse group of heart cells band together to pump blood (which has in it blood cells) so that all the other cells of your body can receive the oxygen and other nutrients they need in order to survive. A diverse group of lung cells band together to allow for the transfer of oxygen in the air to your red blood cells. A diverse group of digestive track cells band together to digest and process your food, take nutrients out of it, and expel the rest.

BioFact

Most cells are microscopic in size, but there are some exceptions to that general rule, including something you might have eaten scrambled this morning. The yolks of eggs are single cells, and they are far from microscopic.

All of these cells have highly specialized purposes, and have differentiated themselves in many ways so that they can accomplish their very specific tasks. But no matter how different they look from the outside, and no matter how specialized they are, they're essentially miniature factories, exquisitely designed to perform their tasks. They're made up of a variety of different parts that perform functions, including copying themselves so that the organ can keep on functioning. And every one of these tasks is guided by the body's DNA, which tells the factory how to go about doing its work.

If you're going to understand how DNA, cloning, and bioengineering work, you first need to understand how cells work. Cells do the actual work of DNA—they carry out the DNA's instructions.

As you'll see a little later in this chapter, cells have inside them a variety of tiny enclosures called *organelles* that perform specific functions. But in order to understand how cells work, it's worthwhile to start looking at something simpler than cells in a body. Instead, we'll look at bacteria, which are composed of only a single cell. We're not alone in first trying to understand bacteria before moving on to understand how a cell works—that's what scientists have done as well. They study the one-celled

BioFact

Even though all the cells in your body have all of your DNA, each cell only uses part of the total DNA. Heart cells only use the DNA they require to do their tasks; muscle cells only use the DNA required to do their tasks and so on. How cells know which parts of DNA to use and which to ignore remains a mystery.

creatures as a way of understanding basic cell biology and functioning, in the hopes that will give them insights into how cells work in higher level animals such as humans.

Let's Start Small: Looking at Bacteria

When it comes to life, you can't get much simpler than bacteria. They're one-celled creatures that don't do anything much beyond growing and then when they get big enough, dividing. Life should be so simple for all of us!

BioDefinition

A **virus** is different than a bacterium—and there is debate over whether it should be considered alive. It's not composed of cells in the same way as living things are; it doesn't have all the cellular machinery that supports life and just exists as a harmless particle until it finds a real living cell to invade. In fact, a virus is really just a chunk of DNA or RNA surrounded by a protective sheath. It can't reproduce by itself. Instead, it hijacks cells of other living things and uses them to create copies of itself, and once those new viruses have been created, they either destroy the invaded cell as a way of breaking free, or else squeeze themselves out through the cell's membrane.

The cell that biologists love to study is E. Coli, which is about one-hundredth the size of a typical human cell. You may have heard of that bacteria because a strain of it has led to outbreaks of disease and even death. But that's the infamous E. Coli *0157*, normal E. Coli's evil twin. There in fact is an entire family of E. Coli bacteria, and most don't do damage—in fact, they're often beneficial. You have countless E. Coli living in your digestive tract, helping you digest your food, for example.

Bacteria are called prokaryotes because they don't have a nucleus, which is a dark organelle in the center of the cell that directs the cell's functioning, and contains the cell's DNA. Bacteria, however, do have DNA; it just isn't isolated in a nucleus. They have a single chromosome curled up into a ball seemingly floating throughout the cell.

In addition to the DNA, a bacteria also has an outer wrapper called the cell membrane, which keeps the molecules needed to sustain life inside the bacteria. Inside the cell membrane is a fluid called cytoplasm, which is made up of about 70 percent water, and 30 percent *enzymes* and other molecules such as amino acids (building blocks of proteins and enzymes) and glucose molecules (a kind of sugar).

As you'll learn in Chapter 4, DNA's job is to make the proteins and enzymes that the bacteria uses in order to live and reproduce.

BioDefinition

Enzymes are kinds of molecules that are vital to the functioning of all life. Their sole purpose is to make chemical reactions occur very quickly. The enzymes themselves aren't changed by the chemical reaction—they're left intact after the reaction takes place. You may have heard of enzymes that help you digest food, but they do much more than that—they're involved in countless biochemical reactions. In fact, the bacterium E. Coli has approximately 1,000 enzymes in its cytoplasm.

Great Things Come in Small Packages?

The cells in plants and animals, and the human body, are more complex than the E. Coli and other bacteria, however. These cells are called *eukaryotes*, or eukaryotic cells because they contain a nucleus. The nucleus contains chromosomes, which hold the cell's DNA. All the cells in a human body have DNA, with one exception—red blood cells. When human red blood cells reach maturity, they lose their nuclei, with all of its genetic material. Because of that, unlike many other types of cells in the body, red blood cells don't duplicate themselves. Instead, red blood cells are created by bone marrow.

While there are differences among all types of cells, most eukaryotic cells have the same basic structure. The nucleus, a dense organelle in the center of the cell, holds the chromosomes and so all the genetic material, and controls all of the cell's functioning. A cell membrane surrounds the cell and protects it as with bacteria. Most of the cell is made up of the cytoplasm. Inside the cytoplasm are a variety of tiny enclosures, called organelles, which perform a variety of functions.

The following figure shows the makeup of a typical animal cell. As you can see, there are some parts that we haven't covered yet. In the rest of this section, we'll take a look at those, and at the nucleus and cell membrane in more detail.

A look inside a typical animal cell.

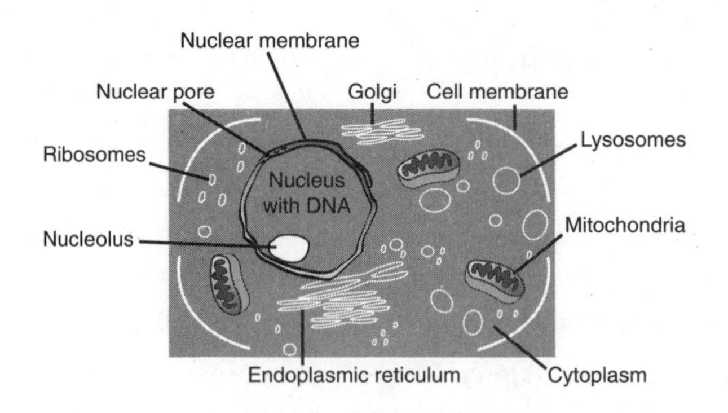

At the Center of It All: The Nucleus

Each cell has only one nucleus, and it directs all the activities of the cell, including ordering proteins and enzymes to be made, and telling the cell how to grow and reproduce. Inside the nucleus are long, twisted strands of material called *chromosomes*, which are composed of DNA and proteins. Along the DNA strands are individual genes. As you'll learn in Chapter 4, the DNA's job is to produce the proteins and enzymes needed for the cell—and the rest of the animal or plant—to survive.

BioDefinition

The **nucleus** is small, but it in turn is made up of smaller units. One part is called the *nucleolus* and this directs the building of the protein factories called *ribosomes*. Another part, the chromatin, contains the chromosomes.

A membrane called *nuclear envelope* or *nuclear membrane* surrounds the nucleus. This membrane is full of tiny holes called *pores* that only allow certain materials to pass in and out of the nucleus. For example, it won't allow the DNA to pass out of the nucleus.

Please Fence Me In: The Cell Membrane

Each cell is surrounded by a membrane that functions somewhat like the skin of a cell. It acts as a selective barrier that only allows certain objects and molecules to pass in and out. So, for example, it keeps the organelles and other vital cell structures inside the cell, and also doesn't allow harmful material in. The membrane allows nutrients to pass into the cell, waste to pass out, and certain proteins and hormones to pass in and out of the cell as needed.

Why Does a Cell Need an E.R.?

Hanging off the nucleus and snaking throughout the cell is a structure called *endoplasmic reticulum* or E.R. for short. It's a convoluted network of tubules and sacs that runs between the nucleus and the cell membrane, and all stops in between.

BioFact

The key to understanding the endoplasmic reticulum is knowing how it gets its name. Endoplasmic means within the cytoplasm, and reticulum is derived from a word that means network. So the word means a network within the cytoplasm.

A lot of work takes place in the E.R.—to a certain extent it's like a factory floor and conveyor belt combined into one. The E.R. is composed of two different regions, the smooth E.R. and the rough E.R. The rough E.R. is dotted with small organelles called *ribosomes*, which is what makes that part of the E.R. rough. The ribosomes do the work of making proteins and enzymes, based on instructions given to it from the DNA in the cell's nucleus. (We'll cover how they do that in Chapter 4.) Smooth E.R. has no ribosomes in it to make it bumpy, hence the name. When the ribosomes make proteins, the proteins travel from the rough E.R. to the smooth E.R., which in turn sends the proteins along its convoluted pathways, frequently to an organelle called the *Golgi apparatus*, which we'll discuss in the next section.

BioFact

The Golgi apparatus was named after the Italian scientist Camillo Golgi (1843–1926) who won the Nobel Prize for Medicine in 1906 for his studies on the structure of the nervous system.

The E.R. performs a lot of other tasks as well, depending on the type of cell. In many cases it adds chemicals like sugars to proteins and enzymes, and is the place where the fatty molecules called *lipids* are made. Although fats have a bad reputation we absolutely need them (for example, the cell membrane is mostly fatty lipids).

More Factory Work

In a field in which strange names rule the day, the organelle we've previously discussed, called the Golgi apparatus, takes the cake. Although it sounds like some kind of Rube Goldberg contraption, it in fact is quite an efficient piece of biomachinery.

After proteins have been manufactured by the ribosomes and sent down the endoplasmic reticulum, they make their way to the Golgi apparatus, which is composed of individual compartments with flattened membranes stacked on top of one another pancake style. Each compartment adds a different kind of sugar, depending on what the protein needs for its job. The Golgi apparatus also does the final "packaging," wrapping the proteins up in tiny droplets of lipid so that they're ready to go out into the cell or the body and start doing their work. So the Golgi apparatus does some final finishing work, and then sends the proteins on their merry way.

Mitochondria: The Cell's Power Plant

Just like a factory needs power in order to keep working, so do cells. All that protein-building and other kinds of activity requires a lot of energy, and the cell needs to get it from somewhere.

That's where mitochondria come in. These organelles are in the cell's cytoplasm, and produce energy so that the factory can keep on chugging away. The more work a cell needs to do, the more mitochondria it needs, and so liver and pancreas cells, which work overtime to help you with digestion, require a whole lot of them, and heart cells, which pump your blood throughout your body every second of the day, need a lot of energy and so are chock-full of them as well.

BioFact

Mitochondria in some ways act as cells within a cell. They divide and duplicate themselves to make more of themselves when they sense that the cell requires more energy. They have their own DNA and ribosomes. In fact, one theory holds that earlier in evolutionary history, mitochondria were originally bacteria that formed a symbiotic relationship with larger cells, and ultimately became part of larger cells, giving up most of their DNA to them.

Mitochondria unlock the energy stored in the sugar glucose and from it produce a molecule called ATP that is kind of like a concentrated energy packet. Those energy packets are then distributed throughout the cell to power the work and chemical reactions that need to be done.

Lysosomes: Taking out the Trash

If you've ever visited a factory, you know that it produces all kinds of wastes and trash. And if the factory is going to keep working properly, it has to dispose of all that in some way.

The same thing is true of cells. With all that protein-building and energy producing, and all the other work that cells do, there can be a lot of waste hanging around. And sometimes bacteria and other foreign matter gets into the cell, and they need to be disposed of as well.

Enter the lysosomes. They're essentially tiny sacks filled with digestive enzymes. They break down and digest things no longer needed in the cell, or that are cell invaders. For example, when mitochondria no longer function properly, lysosomes break them down into smaller parts that can be reused by the cell. They also digest bits of food in the cell. And when the white blood cells of your immune system come across dangerous foreign objects, such as a bacteria, their lysosomes help destroy it.

BioFact

In addition to their day jobs, some lysosomes perform another very vital piece of work, necessary to animal reproduction. When a sperm fertilizes an egg, it needs to bore a hole through the egg's cell membrane so that the sperm can enter. The head of the sperm cell, when it makes contact with an egg, released a lysosome-like material called the acrosome that does the work of boring the hole.

Where Do Plant Cells Fit In?

So far we've talked only about animal cells, but the world is filled with plants as well. Plant cells have the same basic structure and most of the organelles of animal cells, with a few important exceptions. The major difference is that plant cells contain organelles called *chloroplasts*, and without these organelles, there would be no life on Earth. Chloroplasts convert the energy of the sun into food that the plant uses in order to grow and function and they also produce life-giving oxygen. Animals can't create their own food in the way that plants can and need to breathe oxygen, and so all life on Earth is dependent on the chloroplasts doing their job. In some ways, chloroplasts are similar to mitochondria, because they have their own genetic material.

Plant cells also have thick cell walls in addition to cell membranes. These cell walls help support the plant as well as protect the cells. Plant cells typically have more water in them than animal cells, and if they didn't have thick cell walls, the cells would be more likely to burst open.

The final major difference is the presence of a big fluid-filled area called the *vacuole*. The vacuole stores and transports nutrients and waste. They're also used to store substances required by the plant, but which could harm the rest of the cell if they were released. For example, poisonous plants store their poisons in vacuoles. Vacuoles also exist in animal cells but are much smaller.

Where DNA Fits In

So where does DNA fit into this picture? As we've covered, the DNA is located in each cell's command central— the nucleus. The cell's functions are controlled by its DNA. The genes in the DNA oversee the building of the proteins and enzymes necessary for the cells to live, function, fulfill their function in the body, and to reproduce.

BioFact

One mutation that leads to thalassemia is caused by a single genetic "letter" being changed on a single gene. When this letter is changed from a "C" to a "T" the gene can't build the globin protein because the letter change stops the protein-building process part way through the process.

That means that if you control a plant's or animal's DNA, you control what its cells do, and if you control what its cells do, you control that animal. Bioengineer a goat by adding a gene to its DNA that can create the human insulin hormone, for example, and that goat might be able to produce insulin in its milk, which can then be extracted and used by diabetics. The goat can produce the insulin because DNA can instruct its milk-producing cells to manufacture it in their miniature factories.

On the other hand, if a plant or an animal has a mutated gene, its little cellular factories may not be able to produce the right proteins and enzymes, and so diseases may result. For example, a mutation in a gene leads to a disease called *thalassemia*, in which it is difficult or impossible for cells to produce the protein globin. Globin is a component of hemoglobin, which carries blood in our red blood cells, and so it's a serious genetic disorder. (For more information about genetic disorders, turn to Chapter 5.)

It's Not About Sex: How Cells Copy Themselves

Cells don't live forever— they either die or divide into two cells. Each cell is identical to the original cell, including all its DNA. In order for the cell to divide into two, they need enough material to form the two cells.

The most important part of a cell is its DNA found in the chromosomes, and so the most important part of cell division is a process called *mitosis*, in which the chromosomes duplicate themselves so that there are two identical copies of the original set of DNA.

At the beginning of mitosis, the chromosomes duplicate themselves, but the duplicates are attached to one another at a spot called the *centromere*. Next, the chromosomes thicken and become shorter, and they are visible under a microscope as short, squat Xs. (Normally, chromosomes are thin and nearly invisible.) At this point, the chromosomes have duplicated themselves—each half of the X contains the exact same chromosomes. But the duplicate sets are still connected at the middle of the X at the centromere.

Next, the nuclear membrane dissolves, and a fibrous spindle forms. The chromosomes line up along this spindle, waiting for the spindle fibers to pull the chromosomes apart and drag them away from each other, at which point the chromosomes end up at opposite ends of the cell. Each set of chromosomes at each end of the cell is identical to the other. Nuclear membranes form around each set of chromosomes, and the cell gets pinched together in the middle. The two cells break apart, and where there was once only one, there are now two identical cells with the exact same genetic material.

The following figure shows the process of mitosis step-by-step.

Mitosis in a nutshell.

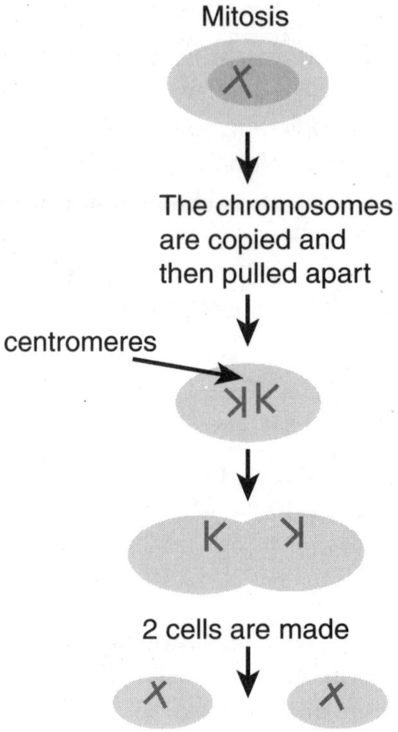

Mitosis

The chromosomes are copied and then pulled apart

centromeres

2 cells are made

Well, Maybe It *Is* About Sex ...

It's all well and good to have two identical copies of the original cell with the complete set of chromosomes in them—in fact it's necessary to life. But that leads to a problem when it comes to reproduction. When animals and plants reproduce, they get one set of genes from each parent. So humans, which have 46 chromosomes, get 23 chromosomes from the father's sperm and 23 from the mother's egg.

But hold on a minute! If each of our cells has 46 chromosomes, and the sperm gives 46 chromosomes and the egg gives 46 chromosomes, that adds up to a total of 92 chromosomes, not 46. And in the next generation there would be 184 chromosomes and so on. Does this have something to do with the new math?

No, the sperm and egg each have only 23 chromosomes, and that has something to do with the way that sex cells are formed, through a process called *meiosis*. As in mitosis, 46 chromosomes start off by doubling and thickening and are joined at the centromere, so that you have two copies of the original genetic material. But then the cells divide not once, but twice, and during that second division the chromosomes don't reproduce themselves. That means that the cell originally started with 46 chromosomes—and then duplicates its chromosomes so that it has 92. When it divides into two cells, each has only 46 chromosomes. And when it divides again, each of the four sex cells ends up with only 23 chromosomes. No new math needed!

Something else important happens during meiosis to form sperm and egg cells. Genes can "cross over" from one set of chromosomes to another. So a gene originally from the mother can cross over onto a chromosome originally from the father and vice versa. This crossing over is important, because it's a way to constantly mix up DNA and make sure that species have a lot of variety in them. This variety is needed for evolution to occur, so that species can adapt to new environments.

What Does All This Have to Do with Cloning?

So what does all this have to do with cloning? Everything. If a new being is created with the exact same genetic material as an existing being, it means that the cells in that new being will have the same instructions as the existing one, and those cells will go about their jobs in the same way. Of course, environment affects animal and plant behavior and so the new being won't act identically to the original because of that, but at the cellular level, the instructions will be the same.

It also points out one of the many potential dangers of cloning. As we've discussed in this chapter, cells divide to produce new cells, or else they die. But cells are programmed to divide a certain number of times, and then after that, problems occur. If you take the genetic material from a cell, and make a clone from it, the new being may be starting off with "old" genetic material that is programmed to start having problems prematurely. That's one of the many dangers of cloning that we'll cover throughout the book.

The Least You Need to Know

- Cells combine into tissues, and tissues combine into organs such as the heart and liver.
- DNA is found on chromosomes in the nucleus of the cell. The DNA has instructions on how the cell can create proteins and enzymes, which is the cell's primary purpose.
- The nuclear membrane surrounds the nucleus and only lets certain molecules through, and so protects the cell's DNA.

- Much of the work of the cell is done within a variety of small enclosures called organelles.
- When more cells are needed they divide by a process called mitosis in which each duplicate cell contains an exact copy of the original cell's chromosomes.
- Sperm cells and egg cells contain half the number of normal chromosomes because they will ultimately combine to give the new plant or animal the correct number of chromosomes.

Chapter 4

How DNA Works

In This Chapter

- A brief history of DNA discoveries
- DNA's double helix structure
- Nucleotide bases explained
- How DNA builds proteins and replicates itself
- What DNA has to do with cloning and bioengineering

At the core of cloning, bioengineering, and genetics is a remarkable molecule, one whose deceptively simple structure has been known for 50 years but whose inner workings are so convoluted that we still don't understand fully how it functions.

That molecule, of course, is DNA. If you're going to understand cloning and bioengineering, you first need to learn about what it looks like, and how it works. As you'll see in this chapter, its code contains answers to the mysteries of life, but there are still many mysteries that have not yet been solved.

What Is DNA?

As we've covered briefly in Chapter 1, *DNA* is a very large molecule that carries the genetic information necessary for all of life. Encoded in it are the instructions for building and maintaining all plants and animals of every size and description on earth.

BioDefinition

DNA stands for deoxyribonucleic acid—a very long name for a very long molecule.

DNA makes up chromosomes, which are found in the nucleus of cells. The nucleus is a kind of central command post for all of the cell's functions. Each chromosome is made up of a single DNA molecule, plus other kinds of matter.

Genes are found along each DNA strand, and these genes literally tell your body how to build itself and how to function. And it doesn't do this by hocus- pocus or magic—it does it in a deceptively simple and mundane way. The purpose of each gene is to create a specific protein or family of proteins, usually an enzyme. And that's it. These proteins do all the work of building plants and animals and instructing them how to function.

Because a DNA molecule is made up of many genes, each DNA molecule contains the instructions for creating many proteins and families of proteins.

One of the science's greatest conundrums is why only certain genes are active in certain cells. Why don't the cells in your eyeballs create proteins used only in your toes? After all cells in your eyeball and cells in your toe contain the very same DNA. Scientists have glimmers of understanding, but don't really know. How specific genes along the DNA molecule are turned on and off throughout your body remains largely a mystery.

From Peas to DNA

Before you learn more about how DNA works, let's take a step back in time. In Chapter 2, we discussed how Mendel began to unlock the mysteries of genetics, and discovered that traits were passed from generation to generation, via something he called genes. His theory gained widespread acceptance eventually, but there was one problem—no one had actually seen a gene.

In the 1920s, a scientist named Fredrick Griffith was experimenting with pneumococcus bacteria that cause pneumonia in mice. He found that although one type of the bacteria caused pneumonia, another type didn't. This type was identical to the disease-causing variety except that it didn't have a specific enzyme for creating a thick outer layer. He killed the disease-causing type by boiling them, and then mixed them with the type that didn't cause pneumonia. He then made an amazing discovery—when the mixture was injected into mice, they died from pneumonia. More amazingly, the type of bacteria found in their bodies was the killer type. What was going on here? He theorized that some kind of transforming factor had traveled from the killer bacteria into the safe bacteria, transforming it into a killer. (He was right—in fact, DNA from the killer bacteria was taken up by the safe bacteria, turning it into the killer strain.)

BioFact

Watson and Crick earned the Nobel Prize for discovering the structure of DNA, but they weren't the only eventual Nobel Prize winners involved in the hunt to solve its riddle. Linus Pauling, who eventually won the Nobel Prize twice, once in 1954 for Chemistry, and once in 1962 for Peace, also tried to understand its structure. However, he thought DNA had a triple helix, like some proteins he had discovered, not a double helix, and so was off the mark.

Later, in 1940, a scientist named Oswald Avery undertook to explain what this transforming factor was. He proposed a molecule known as DNA, but it wasn't until the next decade that scientists confirmed that this was the case. And then, in 1952 James Watson and Francis Crick discovered the actual structure of a DNA molecule, what is now world-famous as the double helix. (A helix is another word for a spiral. So why didn't they call it a double spiral? Which sounds more scientific to you? Enough said.)

A Look Inside the Double Helix

The molecule whose inner structure Watson and Crick decoded is one of the strangest ones in existence. To begin with is its length—a single molecule, if straightened out, would be six feet long, yet when coiled into its original shape, it can fit inside a size measured in the trillionth of inches. And then there's this to consider: There are really only four "working parts" of the molecule, and yet it is responsible for all the varied life on earth.

By the time Watson and Crick came around, it was generally known that DNA was made up in part of a sugar called deoxyribose, a group of chemicals called phosphates, and four bases: Adenine (called A); Cytosine (called C); Guanine (called G); and Thymine (called T). These bases are called nucleotide bases. However, no one knew the molecule's structure or how it all fit together. And discovering that structure was key to understanding how DNA works, because the chemical structure of a molecule often determines its functions.

BioFact

Watson and Crick may have won the Nobel Prize, but in fact, they actually knew very little chemistry. The textbooks of their time had the structure of G slightly wrong. When they tried to build models with G they never worked. A colleague told them about the error and then they had the G models remade. With this and an X-ray picture of DNA taken by Rosalind Franklin, they were in business.

Watson and Crick used scale-model atoms and fit them together to understand the DNA molecule. They noticed that A paired off with T, and that G paired off with C. And they also saw that A would not fit with C, and that G wouldn't fit with T. So they knew that there would be two types of base pairs—*A-T* and *G-C*.

BioDefinition

The **A-T** and **G-C** pairs that hold the double helix together are formed by a force known as hydrogen bonding, a weak attraction between a hydrogen atom on one molecule and a non-hydrogen atom on another molecule.

They proposed that the molecule is a double helix, similar to a spiral staircase. The outside part of the helix—what you might think of as its rails—is made up of the sugar-phosphate combination. And the inside parts of the helix—what you might think of as the steps—are made up of the base pairs, either an A-T pair or a G-C pair. The following figure shows a small portion of DNA.

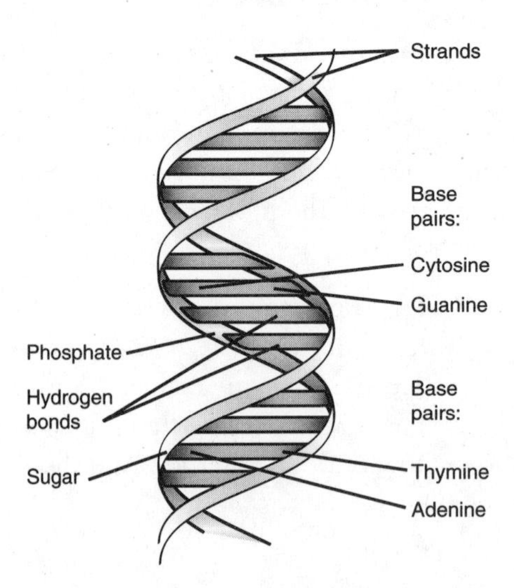

Here's what all the fuss is about—the DNA molecule.

The pairing of A and T and G and C solved something that had puzzled DNA researchers. A scientist named Edwin Chargaff had found that in DNA molecules, the number of As and Ts were always identical, and the number of Gs and Cs were always identical. However, there could be a different number of As and Ts from Gs and Cs. The discovery of the A-T and G-C pairing solved the riddle.

Taking out the Garbage: "Junk" DNA

As you've learned, the DNA molecule contains along its length many genes, which are made up of DNA sequences. Those DNA sequences are used to create a specific protein or family of proteins—a single gene contains the code for a single protein or protein family. But oddly enough, not all the DNA in a gene codes for proteins—some DNA serves no purpose that scientists fully understand, although they know much more about this DNA today than they did in the past. The coding parts of the genes are called exons. The non-coding parts are called by the exceedingly scientific name junk DNA.

There are two types of junk DNA, introns and spacer DNA. Introns sit between exons, separating sections of the gene. Spacer DNA, on the other hand, sits between genes, separating one gene from another. The following figure shows the relationship between exons, introns, and spacer DNA.

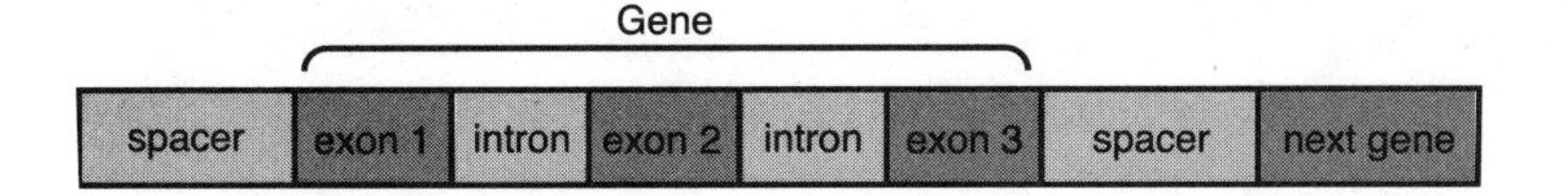

Exons, introns and genes

Scientists believe that introns may be vestiges of long-ago events that combined bits of different genes (exons) into new genes. By making new combinations of exons, new proteins could be made that combined properties of more than one old one. Genes with many exons tend to be more complex and make more complex proteins than those with no exons. So human genes have many exons but bacterial genes usually don't have any.

BioFact

Spacer DNA contains a variety of signals used to communicate with genes. Most of these signals have to do with turning genes on and off in response to factors associated with the cells. It's a mystery, however, why there is so much spacer DNA.

Introns also tell the cells which exons to combine to code for a protein. For example, the exact same gene could make different proteins in your eyeball and in your toe cells. Specific factors in each cell tell the gene which exons to combine to make the protein. That's how a single gene can make a whole family of proteins.

It's Like a Computer for Making Proteins

What have we learned so far? That the entire alphabet of life is spelled out in four letters in the DNA molecule, A, T, C, and G. And that DNA's sole purpose for being is to create proteins. So somehow, DNA is able to build proteins using just those four letters.

Why is building proteins important? Because without proteins, there is no life. Proteins come in many sizes and shapes, and they serve many purposes. Some, for example, are rigid and are the building blocks for tendons and muscles. A protein called *collagen* is a component of skin. Proteins in your eyes are sensitive to light. Others carry oxygen in the blood, or help vital chemical reactions take place in the body, such as digesting food or making hormones. Others are antibodies, which fight disease.

Your organs are made up of proteins as well. While proteins are quite different from one another, they share a basic common characteristics—they are each composed of a single interconnected chain of small building blocks called *amino acids*. Some proteins are small and composed of only a few amino acids. For example, the artificial sweetener aspartame is made up of only two amino acids, phenylalanine and aspartic acid. On the other hand hemoglobin, the protein in red blood cells that picks up oxygen in the lungs and delivers it to the cells throughout your body, is a much larger protein made up of 600 amino acids.

BioFact

There are 20 different kinds of amino acids that make up proteins. Amino acids are made up of carbon, hydrogen, oxygen, nitrogen, and in some cases sulfur. The human body can manufacture many amino acids, but not all of them in large enough quantities, and so these have to be provided in our diet. Amino acids that the body can manufacture in sufficient quantities are called *nonessential amino acids*, and those that the body can't manufacture or can't manufacture in large enough quantities are called *essential*. Nonessential amino acids come either from the food you eat or are made by the "good" bacteria that live inside your gut.

As we've already discussed, a gene's purpose is to create a single protein or protein family—and the larger the gene, the larger the protein it can build, and the smaller the gene, the smaller the protein it can build.

So How Does the DNA Computer Work?

The order of the A, T, C, and G base pair sequences in the gene determines the exact makeup of the protein the gene builds—the protein's shape, size, and the amino acids that it is composed of. In fact, there is a direct relationship between the base pair order and the order of amino acids in the protein that the gene builds. The gene is in essence a blueprint for building the protein. Just like a blueprint, it is not the thing itself—it is not the protein. Instead, the blueprint is copied to make the protein.

As we discussed in Chapter 3, DNA and its genes are located in the nucleus of a cell, rather than in the surrounding cytoplasm. There's a reason for that: Genes need to be protected. If they were floating around in the cytoplasm, some harm could come to them, with potentially disastrous consequences. So they're safe, snug, and secure inside the membrane of the nucleus.

The nucleus of the cell is small and compact, and protein-building generally doesn't take place in there. Instead, it takes place out in the cytoplasm, where proteins are manufactured by molecular machines called *ribosomes*. So we have a problem. The DNA with the blueprint for creating proteins is in the nucleus, and the protein factory is floating around in the cytoplasm. How to get the blueprint to the factory? Simple—send a message that contains a copy of the blueprint.

The messenger that brings a copy of the blueprint to the ribosome factory is called ribonucleic acid (RNA). RNA is a kissing cousin of DNA, but there are a number of differences between them. They both have a sugar-phosphate backbone, and both have bases attached to that backbone. The sugars are different—in RNA it's ribose and in DNA it's deoxyribose. There are also three types of RNA, not just one. RNA usually has a single strand, while DNA has two. RNA is also much shorter than DNA—it can have from fifty to a few thousand nucleotides (remember those As, Cs, Ts and Gs), while DNA can have more than a million. And there's one more difference: While both molecules have As, Cs, and Gs, RNA has a nucleotide called uracil (U) instead of T. The U in RNA pairs with A just like the T pairs with RNA T in DNA.

BioFact

There is only a very small chemical difference between the ribose and deoxyribose sugars—a single oxygen atom. Otherwise, the molecules are the same.

A special kind of RNA called Messenger RNA (mRNA) carries the blueprint for the protein from the DNA in the nucleus out to the ribosomes. Here's how it works. First, an enzyme goes to the DNA, and pulls apart the area around a single gene, in essence unzipping a small piece of the helix and exposing the nucleotides for just that gene. Then the mRNA gets built, using the DNA nucleotides as a blueprint. Each nucleotide in the mRNA is a complement to the corresponding nucleotide in the DNA. So, for example, if there is a G on the DNA, there will be a C on the RNA (remember, G and C are a matched pair), and if there is a C on the DNA, there will be a G on the mRNA. Since mRNA has a U instead of a T, if there is an A on the DNA, there will be a U on the mRNA; and if there is a T on the DNA, there will be an A on the mRNA. Think of it as a kind of funhouse mirror.

BioDefinition

The enzyme that pulls the DNA apart so that mRNA can be copied from the gene is called an RNA polymerase. The process of making the mRNA is called **transcription**.

Let's take an example. The DNA has two strands, one is called the template that does the coding and the other is simply called the non-template strand. Let's say the template DNA sequence is GCGCAATCG. That means the corresponding mRNA sequence will be CGCGUUAGC.

Once this happens, the mRNA is just about ready to deliver its message, but not quite. Remember, genes are made up of DNA that codes for proteins, called exons, and introns, that don't code for proteins. So before the mRNA is sent on its merry way, the intron information is cut out, the other parts of the mRNA stitched together, and it's ready to get a move on.

If that's not complicated enough for you, in real life, it gets even *more* complicated. In fact, through a process known as RNA splicing, a gene can make not just a single protein, but a whole family of proteins. But we don't have time for an encyclopedia-sized dissertation on how that works, so let's move on to the protein-building.

Building the Protein

The mRNA molecule now makes its way out of the nucleus into the cytoplasm of the cell. It's much smaller than DNA, remember, and so can squeeze through the tiny holes in the wall of the nucleus. These holes are lined with proteins called receptors that help the RNA to make it through and direct it to its ultimate home, the ribosome. The ribosome is a small object that looks like a double ball with one ball larger than the other, and made up of approximately 50 proteins and a kind of RNA called *ribosomal RNA*, or *rRNA* for short. The mRNA binds to the ribosome and starts giving it instructions on how to build a protein from amino acids.

Those instructions are contained in sequences that the mRNA copied from the DNA, for example, CGCGUUAGC. The letters are read in groups of three—in this instance there would be three groups: CGC, GUU, and AGC. Each group of three is called a *codon*. A codon represents a specific amino acid. For example, CGC represents the amino acid arginine, GUU represents the amino acid valine, and AGC represents the amino acid serine. A few codons do not represent amino acids. Instead, they tell the ribosome where to start and to stop making a protein.

The ribosome reads the entire sequence from the mRNA between the start and stop codons. It gets the amino acid represented by the first codon, then reads the second codon, and attaches its amino acid to the first one. The ribosome reads the third codon, and attaches its amino acid to the first two amino acids, and so on, building an amino acid chain as it goes along. It eventually comes to one of the "stop" codons that tells it to stop building the protein. At that point, the ribosome has finished its manufacturing job. The chain of amino acids it has built is a completed protein.

BioFact

There are a total of 64 codons, but only 20 amino acids. So what gives? Has someone made a mistake and done the math wrong? How can each codon represent an amino acid—the numbers don't add up? Codons can actually be "synonyms," that is, several codons can represent the same amino acid. So, for example, the codons CGU, CGC, CGA and CGG all represent the same amino acid, arginine.

The protein floats loose from the ribosome, and it folds into a very specific, very complicated shape. This precise shape of a protein is vital to its functioning—only when it is folded properly can it do the work it was designed to do. The way that proteins fold is not completely understood, and is remarkably complex, so much so that supercomputers are sometimes devoted to the task of understanding how a certain protein takes its ultimate shape.

In some instances, the protein is now finished and can now go off and get to work—building a brain cell, helping you digest food, or whatever else its function is. In other instances, though, a bit more work needs to be done to it. In some instances, a small piece needs to be cut out of the protein. In other instances, sugars or lipids (fats) or phosphates or metals have to be added to the amino acid chain. For example, interferon, a protein that helps the body fight off diseases, needs to have specific sugars added to the chain in order to work.

It's Off to Work We Go

Finally, the DNA has done its work and has built a shiny new protein straight off the assembly line. But many proteins need to have lots of those sugars or lipids added and they need to go to different places to do their work. For those proteins, yet one more step needs to be taken. They are sent to the Golgi apparatus, which you read about in Chapter 3. The Golgi acts as a kind of transportation center—it packages the proteins with sugars or lipids and then sends them off to their final destination. Some proteins are sent outside the cell and others are shunted to places inside the cell where they do their job.

This whole process of building proteins from DNA sequences can be very confusing. Figure 4.3 should help you understand how it all fits together.

Here's how proteins get built from DNA.

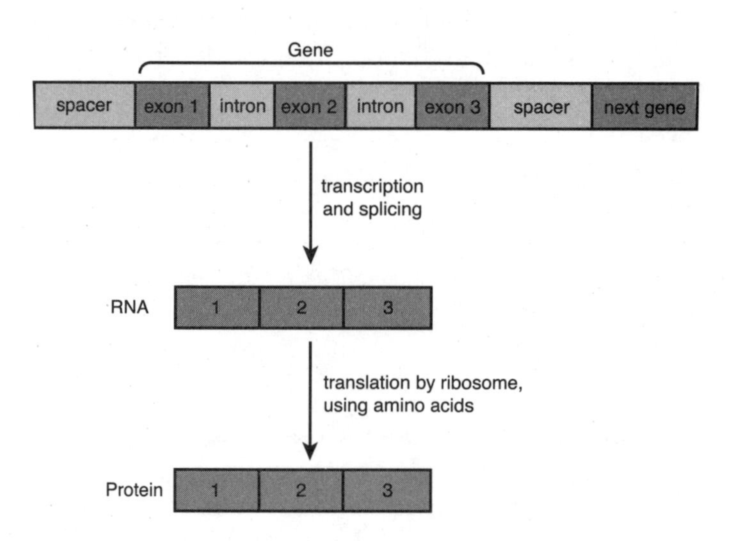

A Mystery at the Core of Life

All this makes perfect sense and is relatively well understood. But there's a mystery that scientists still don't understand, one at the very core of life. Every cell in our body contains all of our DNA, and yet each cell only creates very few specific proteins. For example, the cells in your eyeballs don't create the enzymes necessary for digesting your food—if they did, we'd all be stuffing pizza up our eyes. And the cells in your lungs don't create the proteins needed to build your fingernails—unless, of course, you're called leather-lungs by your friends. Yet every cell contains all of the 30,000 or so genes in our bodies. That means that some unknown mechanism tells the cells in your eyeballs to turn on only the genes that make proteins your eyeballs need, and tells the cells in your lungs to turn on only the genes that make the proteins that your lungs need. All the other genes are switched off.

Breaking up Isn't Hard to Do

DNA does one more important thing in addition to building proteins—it copies itself when cells divide. Whenever a cell divides in two, each of the two cells needs to have the exact same chromosomes and DNA that were in the original cell. If you're mathematically minded, you've no doubt noticed a problem here: How can one set of chromosomes and DNA go into two cells? The answer, of course, is that the DNA replicates itself, and each cell gets one full set of the original DNA.

That's also how DNA gets passed from generation to generation—copies of the DNA in the sperm and egg cells are passed from one generation to the next.

As we discussed in Chapter 3, when cells divide in the process called mitosis, the chromosomes first duplicate themselves, and are then carried over into the two new cells when the process is complete. Each new cell contains the entire set of chromosomes of the original parent cell—all of the 30,000 genes.

BioDefinition

Sperm and egg cells form in a slightly different way than all other cells in the body, through a process called **meiosis.** A master "germ" cell divides to produce four sex cells. Each new cell has half the DNA content of the master. They have half the number of chromosomes because when sperm and egg combine, it will make a full set. If each cell instead had the full number, when the cells combined they would have twice the number of chromosomes.

In theory, DNA replication is exceedingly simple. Each of the two strands of the DNA's double helix has the information necessary to create its complementary strand. That's because Gs always line up with Cs, and As always line up with Ts. So if you pulled the strands apart, each could only be copied in one way and the result would be two copies of the original double helix. Check out the following figure to see how it works.

Of course, real life is always much more complicated than theories, and so the actual way that DNA replicates takes more steps than this—and in fact, the whole process isn't yet fully understood. First, enzymes force the two DNA strands apart at a small area of the DNA called the origin. As the molecule is pulled apart, each of the two resulting strands attract the proper nucleotides needed to make a copy of it. If an A is exposed, it attracts a T to it, but repels Gs and Cs. As the separation proceeds, each strand is building its complementary strand.

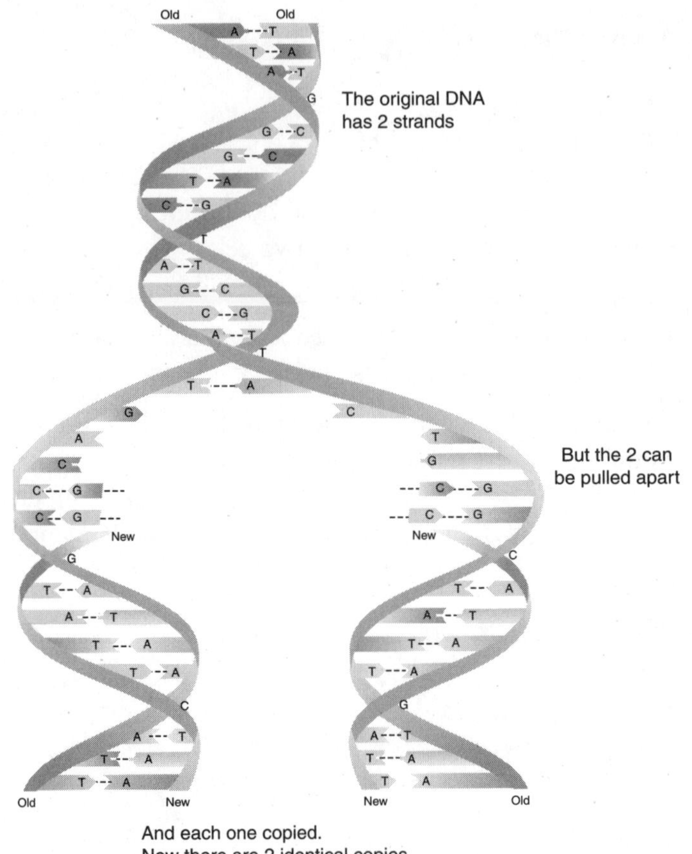
Old
Old
The original DNA
has 2 strands
But the 2 can
be pulled apart
New
New
Old
New
New
Old
And each one copied.
Now there are 2 identical copies

The simple mystery of DNA replication.

Danger for the Clones!

One aspect of the way that DNA replicates itself has significant health consequences for clones, and may lead to health problems with cloned animals, notably premature aging. (For more information about health problems associated with clones, turn to Chapter 6.)

The ends of chromosomes are protected by structures called *telomeres*. Every time chromosomes divide, the telomeres shorten slightly. Over time, they shorten more and more until they are no longer capable of functioning, and the cell dies. Think of them as a kind of biological clock counting down toward the time a cell dies.

The DNA used for cloning comes from a cell that has divided many times already and so already has shortened telomeres. Shortened telomeres are thought to lead to numerous problems, most prominently premature aging. So clones may start out biochemically older than noncloned animals, and may live shorter lives.

Of Mutants and Men

As you read this, tens of thousands of cells in your body are dividing, replicating their DNA—in fact about 100,000 cells divide every second. Each of those 100,000 cells has to replicate all of its DNA, which means about 30,000 genes each. Each of those genes has a specific chemical structure and arrangement of nucleotide bases (the pairs of As, Ts, Cs and Gs) that have to be copied exactly—there are, in fact billions of nucleotide bases in your genes. If you do the math, that means that every second of your life, literally hundreds of trillions of nucleotide bases are being copied. So multiply that number by the number of seconds that you're alive .... Actually, don't bother. As you might guess, the number is beyond astronomical.

Considering the number of bases that have to be copied, it's only natural that on occasion a mistake can be made. Perhaps an A gets put in place instead of a G, or a T instead of a C. Your body has ways of fixing those errors. Enzymes look for As, Ts, Cs, and Gs in the wrong places and then cut them out, and other enzymes find the right As, Ts, Cs, and Gs and put them in their proper places.

Sometimes, though, the mistake makes it through, and the new gene isn't an exact copy of the original gene. In that instance, the change is called a mutation. Often, the mutation doesn't really matter; it creates a different protein than the original gene, but that new protein does no harm. And still other times, the mutation produces a protein that can do harm. You'll learn more about mutations in Chapter 5.

> CAUTION
>
> **BioWarning**
>
> Mutations in your genes don't only occur naturally in the body—outside forces can cause mutations as well. Ultraviolet light from the sun can cause DNA mutations, as can radioactivity and certain chemicals.

Cloning, Bioengineering, and DNA

So what does this all have to do with cloning and bioengineering? Everything. When you clone an animal or plant, you're making a new copy of it with the exact same makeup of DNA as the original animal or plant. That means that every gene is exactly the same as the original and so it makes the exact same proteins as the original plant or animal. That's why clones look exactly like the original. And as you'll learn in Chapter 6 some cloning requires that you turn on or off specific genes during the cloning process in order to make it all work.

As for bioengineering, it means even more. A specific gene produces a specific protein, no matter what animal or plant it is in. If a gene produces a peanut protein, it will produce that protein whether it is in a peanut plant or a soybean (with certain exceptions, that is). That opens the way for bioengineering of all kinds. Let's say, for example, that you want to produce large amounts of human insulin, so that diabetics can have enough of it to keep their diabetes under control. Insulin is a protein created by a specific human gene. So if you could somehow get that gene into an animal in a certain way, you could have that animal produce large amounts of insulin in its milk, and you could then harvest that insulin for people to use.

Or, for example, you might want rice to have a higher protein content, or have more vitamins than it normally has, so that people in poverty-stricken regions could grow that rice and not suffer from malnutrition. Those proteins and vitamins are produced by specific genes, and so if you could put those genes into rice, the rice would help fight malnutrition. These are, in fact, two of the ways that bioengineering is being used today. You'll learn more about it in Part 4.

The Least You Need to Know

- DNA makes up chromosomes, which are found in the nucleus of cells. Many genes are found on a single strand of DNA.
- DNA looks like a double helix—something like a spiral staircase. The outside portions are made up of sugars and phosphates, and the inside "steps" are made up of the nucleotide bases known as A, T, C, and G.
- A single gene produces a single protein or family of proteins.
- Proteins are made up of two or more amino acids. There are 20 common amino acids.
- Messenger RNA copies the structure of a gene and carries it from the nucleus of the cell out to the cytoplasm, where ribosomes read the gene's structure and assemble amino acids into proteins based on what it finds.
- DNA replicates itself by splitting into two pieces. Each piece assembles the proper nucleotide bases to make an exact copy of the original.

Chapter 5

How Genes Build Plants, Animals, and Humans

In This Chapter

- What is a genome?
- What is a proteome?
- About the Human Genome Project
- How mutations cause genetic disease
- How gene therapy can cure disease

Looked at one way, all life is just a series of genetic letters—the nucleotide bases that make up DNA. But how those letters are arranged means everything to life—they make a man a man, a mouse a mouse, and a mustard weed a mustard weed.

As you'll see in this chapter, what ultimately determines what an organism is and how it functions is its genome, the sum total of its DNA. So in this chapter we'll look at the human genome, at how it creates organisms, and at the consequences of when things go wrong in the genome.

Of Mice and Men

Before we can take a closer look at how genes build plants, animals, and human beings, let's take a step back for a moment and review what you've learned so far about DNA and genes:

- All living things, whether they be mice, men, or mustard greens, have the same basic material that determines how they live and grow, and that controls all their biological processes—DNA.
- DNA makes up a living being's chromosomes, and is passed down from generation to generation. It is a very large, complex double-helix-shaped molecule, and is composed of four nucleotide bases (called in shorthand As, Cs, Ts, and Gs), a sugar called *deoxyribose*, and a negatively charged chemical called *phosphate*.
- Along the DNA strand in a chromosome are many genes, the individual units of heredity.
- The sole purpose of genes is to create proteins and enzymes. Each gene creates a single protein or enzyme or family of proteins or enzymes.
- The exact arrangement of the nucleotide bases on the DNA in genes determines what proteins or enzymes those genes will create.
- Proteins and enzymes interact with each other in very complex ways, and ultimately, their interaction is what builds and sustains life.

What Is a Genome?

The sum total of all the DNA in the chromosomes of a particular organism is called its *genome*. Every cell in an organism, with a few exceptions, contains the organism's entire genome. Even though each cell contains the entire genome, most of the genes in a given cell don't actually produce proteins or families of proteins—an eyeball cell, for example, doesn't produce digestive enzymes. Scientists are increasingly knowledgeable about which genes turn on and off in which cells, but much of how this occurs remains a mystery.

As a general rule, the more complex the organism, the more genes it has, although that's not always the case. Men and mice, for example, each have about 30,000 total genes, and Mickey and Minnie Mouse aside, humans are far more complex beings than mice. When was the last time, for example, you saw a mouse hit a baseball, read a book, or cook a flank steak? On the other hand, humans aren't that great at chewing through baseboards.

BioFact

The only cell in the human body that doesn't contain the complete human genome (except sperm and egg cells) is the mature red blood cell. It contains no nucleus, and therefore has no genome. Red blood cells start out with nuclei and genes, but as they mature, they lose their nuclei and fill up with hemoglobin. This protein carries oxygen throughout the body. Because the mature cells have no nuclei, they don't replicate themselves as do the other cells in the body. Instead, as the old red blood cells die they are replaced with new red blood cells that are created in the bone marrow.

So it's not just the total number of genes in an organism's genome that determines the complexity of that organism. It's also the size of individual genes, and how those genes create protein and enzyme families. Human beings, for example, can create many related proteins from a single gene—a protein family—using a process called differential splicing, while bacteria can only create a single protein from a single gene.

Many people think of a genome as the sum total of genes in an organism, but in fact, that's not quite true. The genome contains the organism's DNA, but much of this DNA is not part of genes. In Chapter 4, we mentioned that chromosomes contain a great deal of so-called "junk DNA" that in fact is not junk at all. Rather, it is intricately involved in the process by which DNA builds proteins and protein families. So the genome is composed not only of genes, but of all the DNA found outside of genes on chromosomes as well.

Measuring the Size of a Genome

Scientists measure the size of a genome in two different ways—by the total number of genes in the genome, and the total number of nucleotide bases. So, for example, the human genome has approximately 30,000 genes and 3 billion bases, while the laboratory mouse has approximately 30,000 genes and 2.6 billion bases. The following table shows the genome size of a variety of organisms. As you can see, we're at the top of the heap.

BioFact

The average gene in the human body consists of 3,000 bases, but there is a tremendous variation in the size of genes, with some much smaller and some much larger. The largest known human gene is dystrophin, which is composed of 2.4 million bases. A defect on this gene, which is found on the X chromosome, can lead to muscular dystrophy.

The Size of the Genome of Various Organisms

Organism	Size of Genome (number of bases)	Estimated Number of Genes
Human being	3 billion	30,000
Laboratory mouse	2.6 billion	30,000
Mustard weed	100 million	25,000
Roundworm	97 million	19,000
Fruit fly	137 million	13,000
Yeast	12.1 million	6,000
E. Coli bacteria	4.6 million	3,200
HIV	9,700	9

On to the Proteome

As we've told you several times in this book so far (and no doubt, will tell you several times more), proteins are what build and maintain life. Each individual cell in the body has in it many different proteins, and the constellation of all the proteins in a cell is called its proteome. The study of the proteome is called *proteomics*. The proteome of a cell, unlike the genome of an organism, is dynamic, literally changing from one moment to the next, in response to tens of thousands of changing signals from inside and outside the cell. Each protein's behavior and makeup is specified not only by a gene sequence, but is also affected by its relationship to and reaction with all the other proteins in the cell.

BioDefinition

Proteomics is the study of all of the cell's protein structures and activities. Because proteins and their activities underlie all of life's functions, the study is becoming increasingly prominent, since it holds the key to understanding human health and disease.

All About the Human Genome Project

Scientists are an ambitious lot. When looking at the sheer size of the human genome—some 3 billion bases—the nonscientist may look at his watch, notice it was just about nap time, and say he'd get to it tomorrow. And that tomorrow may never come.

But scientists are made of sterner stuff. They look at their watches and think, "Time to get started."

Back in 1990, the federal government, spurred into action by many scientists, announced that it was going to crack the genome—it launched an effort to map every single gene and also learn the sequence of nucleotide bases in the human genome. And the government, with its flair for the obvious, called the effort the Human Genome Project.

It would take 15 years and several billion dollars to accomplish the task, it announced. Cynics back then may have used the unofficial Government Productivity Index as a means of calculating when real results would come in, and figured that in about 30 years things might finally work out.

For once, the cynics were wrong. The government was smart and farmed out the work to universities and research institutes. Ten years later, in June 2000, the first working draft of the map of the human genome was completed. In February 2001, the analyses of the details of the map were published in the scientific journals *Nature* and *Science*. And then in April 2003, the so-called high-quality reference sequence was finished, the more complete and polished work. The labor was done, two years ahead of schedule.

BioFact

Learning the sequence of the Human Genome was not always a cooperative project—it in fact was fraught with controversy. Two separate groups eventually raced to map the genome, a government-funded consortium funded by the National Institutes of Health (NIH), and a group under the private biotechnology company Celera Genomics. Part of the disagreement arose because Celera planned to use information for profit, while the NIH planned to make all the information available for free to all. But the groups also disagreed about the methods that should be used to map the genome. That's one reason why the map was published in two separate scientific journals rather than one—each group published in a different journal. In the end, the project was ultimately a success, despite the friction.

What Does It Mean to Map and Sequence the Genome?

All that may sound very impressive, but you may be wondering, what exactly does it mean to map and sequence the human genome?

First, one sometimes creates a map of every gene on each chromosome. So for each individual chromosome, this map shows the exact location of each gene, and where each gene is in relation to every other gene. In that way, it's a lot like a roadmap.

But in the end one needs to determine the exact sequence of the base nucleotides (As, Cs, Ts, and Gs) on all the DNA in the genome. So it means not only knowing the location of each gene, but also knowing the precise sequences of As, Cs, Ts, and Gs for each gene as well, as well as for the DNA between genes—and that means the exact order of an astonishing number of bases, some 3 billion in all, give or take a few.

BioFact

Human chromosomes vary in size a good deal. The largest chromosome is Chromosome 1, and has 263 million bases on it. The smallest, Chromosome 21, has 50 million.

Sounds tough? Well, the project wasn't content just to take that on. Other goals included understanding gene functions through making genetic comparisons between mouse and men, creating better computing resources for supporting future genetic research and commercial applications of research, studying human genetic variation, and training future scientists in genomics.

Oh, and in their spare time, they decided to map the genome of a number of other organisms as well, including mice, yeast, bacteria, and others.

Why Bother Mapping the Genome?

The Human Genome Project took more than a dozen years and billions of dollars to complete. But why bother to spend all that time and money—was this really just a giant welfare project for scientists?

In fact, the payoff will most likely prove to be tremendous, and will most likely lead to an enormous number of health and scientific benefits. DNA has something to do with almost every aspect of human health and disease, so understanding how DNA and gene sequences work is expected to dramatically affect how disease can be prevented, diagnosed, and treated. In fact, it will be one of the most significant advances in public health in history.

If that's not enough, here are some benefits in a little more detail:

- **Gene testing** The project will greatly expand the potential for gene tests, which can be used to diagnose disease, provide a prognosis for diseases, and predict risks for diseases in healthy people and their offspring or potential offspring.
- **Designer drugs and therapies** Rather than current "one-size-fits-all" methods of drug therapy, drugs could potentially be customized to a person's particular genetic makeup or the specific nature of their disease.
- **More effective drug development** It will lead to the development of more highly targeted drugs aimed at specific biochemical events that lead to disease.

- **Gene therapies** Genes can be used to treat disease—for example, to essentially implant a healthy gene into the body of someone who has a mutated gene that leads to disease. That "healthy" gene can then produce the protein or family of proteins that the body needs in order to function normally. We'll look at this in more detail in Chapter 22.

BioFact

Since everyone's genome is different, the question naturally arises: whose genome was mapped by the Human Genome Project? The project gathered DNA samples from a large number of donors, and used only a few of them as sources, so the genome is a composite of multiple people's. In fact, Craig Venter, who headed up Celera's portion of the project, used his own DNA as one of the five samples his group used. He claims that as a result of examining his own sequence, he is taking cholesterol-lowering drugs.

- **Bioarcheology** By comparing the variations in the genes of individuals in geographically dispersed areas, it is possible to learn how groups have changed and migrated throughout human history—for example, tracing the history of the Jews after the Diaspora when the Jewish Temple was destroyed in Jerusalem by the Romans. DNA has even been extracted from mummies to learn about the Egyptian pharaohs.
- **DNA forensics** Through the development of DNA "fingerprints," it has already become easier to identify individuals more accurately from minute samples of their DNA. All that's needed is a little saliva or hair.

How the Genome Was Mapped

In order to map the human genome, the exact sequence of base nucleotides had to be determined. It's a very complex process, and so we can't go into all the details here, but there were two different approaches taken to the problem—one by each of the groups involved in mapping the genome. In the NIH method, chromosomes were first divided into shorter pieces—a necessity, considering that each chromosome has from 50 million to 250 million bases in it. These pieces are "mapped" so that their order along the chromosome is known.

Each of those short pieces is then copied, base by base. The scientists can then learn which of the four bases is needed to make the copy grow. In this way they learn the order of bases in each piece. This then identifies the sequence of As, Ts, Cs, and Gs

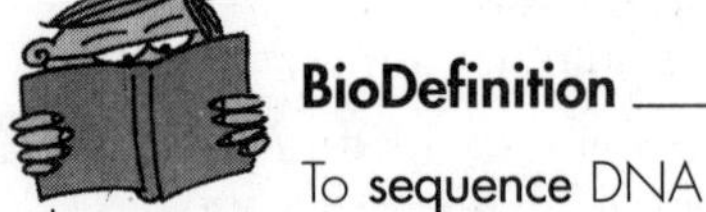

BioDefinition

To **sequence** DNA means to figure out the exact order of its nucleotide bases (As, Cs, Ts, and Gs).

in the short pieces and the short pieces are placed in the order shown on the map. Voilà—a genome *sequence*.

In the Celera method, more of a shotgun approach was used. The genome was broken up into millions of random fragments, and each individual fragment was sequenced. Then powerful computers stitched everything together and figured out where they belonged on the genome.

Obviously, all this is an incredible simplification—otherwise, why spend billions of dollars and take more than a dozen years in what can be described in a few paragraphs?

Some of the Project's Findings

The project has discovered an enormous amount of information—as we've mentioned before, it found out that the human genome has 3 billion nucleotide bases and that the average gene contains 3,000 bases, but that they vary greatly in size, with the largest having 2.4 million bases. Here are some other facts it discovered:

- The human genome sequence is almost identical in all people—it is 99.9 percent the same.
- Only 2 percent of the genome contains the code with instructions for synthesizing proteins. The rest of the DNA serves a variety of other purposes.
- So-called "junk DNA" (the DNA between genes that scientists now realize perform useful functions) makes up at least 50 percent of the human genome. By way of contrast, mustard weed contains 11 percent junk DNA; the roundworm contains 7 percent junk DNA; and the fruit fly contains 3 percent junk DNA.
- Genes are concentrated in what look like random areas of the human genome, and between them are huge areas that contain DNA that does not code for proteins.
- Other organisms' genome is often more uniform than the human genome, with genes spaced more evenly throughout the genome.
- Chromosome 1 has the most genes, 2,968, and the Y chromosome has the fewest genes, 231.

- More than 40 percent of human proteins share similarities with fruit-fly or worm proteins, but humans have three times as many kinds of proteins as the fruit fly or worm because of the way that a single gene can create multiple proteins.
- Human sperm cells have about twice the number of mutations as egg cells. Scientists believe there are a number of reasons for this, including that it takes more cell divisions to create a sperm than to create an egg.

> **BioSource**
>
> The United States Department of Energy, which funds a great deal of genome research, has an amazing interactive website that lets you click on any human chromosome, zoom in, and then see a list of genetic disease associated with genes on that chromosome. Find it at: www.ornl.gov/TechResources/Human_Genome/posters/chromosome/chooser.html.

These are just a few of the findings. And each time scientists find out something new, it raises new questions. As the project's scientists concluded in their paper, "the more we learn about the human genome, the more there is to explore."

Take a Quick Tour of the Map

The map of the genome is publicly available, and so anyone can view it on the Web. There are a number of maps available, including one from the issue of the journal *Science* which was one of the two publications that published the initial map, and another from the federal National Center for Biotechnology Information.

The map from *Science* is of more interest to the nonscientist, because in addition to showing information about each chromosome, it also highlights particular genes, and the consequences of mutations in those genes. To get to the map, visit www.ncbi.nlm.nih.gov/SCIENCE96/.

A map published by the International RH Mapping Consortium is a more up-to-date and complete map, and so is of more use to scientists, but is of less interest to laymen. Get there by going to www.ncbi.nlm.nih.gov/genemap99/.

Genetic Diseases, Genes, and Proteins

A great deal of the work that follows the mapping of the human genome is understanding what the mapped genes do. Just because we know the genes' sequences doesn't mean that we understand what the proteins and protein families the genes code for actually accomplish in the human body.

However, we increasingly know what individual genes do, and how their mutations can cause genetic diseases. Typically, when a gene mutates, disease is caused because the protein or protein family the gene encodes for either cannot be produced at all, or more often, is produced incorrectly. The loss of the proper function of these proteins is what causes the disease. In some instances, more than one gene is involved.

BioFact

Among the strangest and rarest of genetic disorders is fish-odor syndrome, which occurs because of a mutation on a gene found on Chromosome 1. In fish-odor syndrome, a person literally smells like a fish. Also called *trimethylaminuria*, it is caused by a genetic mutation that stops the body from producing an enzyme called FMO3. Because of this, the liver cannot metabolize a chemical called *trimethylamine*, and the chemical is excreted in sweat, urine, and breath, causing a very strong fishy odor. There have been only about 200 to 300 documented cases of the syndrome.

There are several different kinds of mutations that can occur. There can be a "misspelling" in the gene, so that one or more of the As, Ts, Cs, and Gs is changed into a different one. Sometimes there can be extra DNA in a gene. Or pieces of DNA can be lost. And sometimes the DNA gets shifted around within the chromosomes. Some genes mutate more frequently than others. Scientists call these genetic hot spots.

About Point Mutations

The simplest kind of mutation is a point mutation, in which there is a single misspelling of the gene's sequence of As, Ts, Cs, and Gs. Sickle-cell anemia, in which red blood cells are shaped like a sickle, is usually caused by a mutation like this. It can damage blood vessels, internal organs, and lead to brain damage and heart failure.

BioFact

Not all genetic disease is caused by recessive genes. Some are caused when only one of the genes is mutated. An example is Huntington's Disease, a disorder of the central nervous system that typically appears in someone between the ages of 30 and 50. This is the disease that killed the famous folk singer Woody Guthrie.

As with many genetic diseases, sickle-cell anemia is only caused when a child gets two copies of the mutant gene, one from each parent. It is recessive, so if one normal gene is present, the disease normally won't occur.

The Case of the Missing DNA

In some mutations, one or more coding parts of a gene are missing, and so an incomplete protein, or no protein at all can be created. An example of disease caused by this is cystic fibrosis, in which a great deal of mucus accumulates in the lungs.

Hundreds of different types of mutations can cause cystic fibrosis. But in the most common type of mutation leading to the disease, only three genetic letters are missing.

Too Many Letters Lead to Trouble

In another kind of mutation, extra As, Ts, Cs, and Gs are put inside the gene, which disrupts the normal functioning of the gene. Elephant Man Disease, or *neurofibromatosis*, is caused when noncoding repetitions of DNA sequences are put into a gene. And Huntington's disease is caused when the sequence CAG is repeated numerous times in a gene.

Too Many Chromosomes

In some cases, a mutation affects an entire chromosome, not an individual gene or several genes. This is a very severe kind of mutation, and so frequently an embryo with this kind of mutation will abort spontaneously.

However, that doesn't always happen. The most common severe chromosome mutation is Down Syndrome, in which a person is born with an extra copy of Chromosome 21, and so has three copies of it, rather than the normal two. This leads to mental retardation as well as a variety of distinct physical characteristics. Another disease of this sort is *fragile X syndrome* where the X chromosome is broken.

Genetic Malfunctions Don't Always Leads to Disease

In many cases, a genetic malfunction will not necessarily cause a disease—certain environmental conditions, combined with a mutation, cause the disease. That means that people may have the tendency toward disease—toward high blood pressure or toward developing breast cancer, for example—but that disease will only develop when certain conditions are met.

Where Gene Therapy Fits In

Many scientists believe that gene therapy will be able to cure many diseases that are created by mutant genes. In gene therapy, a healthy gene is inserted into cells and taken up by them, and so the gene functions properly by coding for the right protein or family of proteins. However, it is still in the experimental stage, and has proved to be controversial because in some cases, people who have undergone the therapy have developed cancer (see Chapter 22).

The Least You Need to Know

- The sum total of all the DNA in a particular organism is called its genome. Every cell in an organism, with a few exceptions, contains the organism's entire genome.
- As a general rule, the more complex the organism, the larger its genome. The human genome has 3 billion bases and 30,000 genes, while yeast has 12.1 million bases and 6,000 genes.
- The constellation of all the proteins in a cell is called its proteome. The proteome of a cell, unlike the genome of an organism, is dynamic, literally changing from one moment to the next.
- The Human Genome Project mapped and sequenced the entire human genome. It cost billions of dollars and took 13 years to complete.
- When a gene mutates, it can cause disease because it can't properly code a protein or family of proteins.
- In gene therapy, healthy genes are inserted into cells to replace mutated ones. The therapy has been controversial, because a number of people who have been cured by the therapy have later developed cancer.

Part 2

How Cloning Works

What, exactly is a clone? How are plants and animals cloned? How can cloning help humankind? Can human beings really be cloned?

And most important of all, why in the world was the first cloned mammal, Dolly the sheep, named Dolly? (Hint, it has to do with a world-famous country singer and her frequent choice of revealing clothing.)

In this part you'll learn all you need to know—and more—about the science of cloning and its potential benefits. But don't worry—it's not hard to understand and by the end of it you won't be seeing double.

Chapter 6

What, Exactly, Is a Clone?

In This Chapter

- Understanding what a clone is
- How plants are cloned
- How animals are cloned
- How Dolly the sheep was cloned
- What problems are there with cloning?

You've heard about cloning, you've read newspaper articles about cloning, and you've probably seen a science fiction movie or two in which cloning played a factor as well. So you're probably thinking you're an expert on the subject.

You may well be, but the odds are that you're not. Most people only have the haziest idea of what a clone actually is, or how one goes about creating a clone. So in this chapter we'll cover what clones are, how they're created, and some of the potential problems with clones.

Seeing Double: What Are Clones?

Aldous Huxley, in his 1958 novel *Brave New World*, presented a chilling vision of a society populated by the results of mass cloning—and, in fact, a society devoted to cloning human beings as a way to control behavior and action. Mass centers were devoted to the task, such as the Central London Hatchery and Conditioning Centre. Up to 17,000 copies of human beings at a time form the same genetic material.

Some 15 years later, Woody Allen weighed in with his own vision of cloning. In *Sleeper*, Mile Monroe, a mild-mannered health store owner from New York's Greenwich Village, is cryogenically frozen and wakes up 200 years later, where he finds himself in the middle of a revolution. The tyrant who he is trying to overthrow is dead; all that remains is a nose that will be cloned to make a new leader—and to make an escape, Monroe holds the nose hostage, threatening to shoot it unless he is set free.

Neither man quite got the science right. What do you expect? Huxley was a visionary, writing well before scientific breakthroughs that make animal cloning possible—before, even, the DNA molecule was decoded. And Woody Allen … well, what is there to say? It's best not to look to our great comedians for the truth about science.

Although their science may not have been quite right, they have expressed many people's attitudes about clones and cloning. There's a lot of misunderstanding about what clones are—and we'll try to dispel them throughout this chapter.

So what is a clone? Put simply, it's a copy of another organism that has the exact same DNA as the original.

Cloning is nothing new—nature has been doing it for millions of years. When plants like strawberry plants and potatoes send out runners, modified versions of a stem, new plants grow wherever the runners take root. Each new plant is a clone of the original.

BioFact

The San Diego Zoo has become a kind of modern day Noah's Ark by the use of cloning. The so-called Frozen Zoo is a collection of frozen tissue samples from endangered species from around the world, such as pandas, gray whales, and condors. If any of those species becomes extinct, the thinking goes, they will be able to be cloned using the frozen tissue, because cells in the tissue contain the necessary DNA for cloning. It's more than a theory: Scientists were able to clone bantengs, a kind of rare, endangered cattle from the island of Java, using frozen tissue from a banteng that had died 20 years earlier. Scientists inserted Banteng nuclei into enucleated cow eggs and transferred the resulting embryos to cows.

Certain animals can clone themselves. A tiny water animal called the hydra can clone an entire new identical hydra from the original when a small part is cut off. That small part grows an entire new hydra with the exact same genetic material as the original.

Even higher-level animals, under certain conditions, can to a certain extent clone themselves through a process called *parthenogenesis*. Some animals, such as certain insects, worms, and some species of lizards, fish, and frogs, can develop into adults from unfertilized eggs, in certain environments. Each of the animals that develops will get genetic material only from the mother, and none from the father.

The most obvious example of cloning is human identical twins. As you learned in Chapter 2, when an egg is fertilized by a sperm, the new cell gets half of its chromosomes from the sperm and half from the egg. That fertilized egg begins to divide and all the cells have the exact same chromosomes as the original egg. It turns into an embryo, and the embryo then develops into a person.

However, if the fertilized egg splits into two identical cells, each of those cells can develop into an embryo and turn into a person. Because the two initial cells have the identical DNA, the resulting twins have the identical DNA. The result: identical twins.

BioFact

Identical twins in humans are clones of one another—they come from the same egg that splits, and each of the two resulting embryos has the same genetic material. Fraternal twins, however, are not clones, and can be as different as any other kind of sibling. For example, one can be a boy and one a girl. Fraternal twins occur when two different eggs are fertilized by two different sperm, and each embryo develops separately. The embryos have different genetic material, and so are not clones.

How Cloning Works

Nature has been busy cloning organisms for a long time, but only in more recent times has man gotten into the act. The big breakthrough, of course, was when Dolly the sheep was cloned in 1997. But we have been cloning plants and animals for a long time before then—in fact, when it comes to plants, we've been cloning for thousands of years. Much agriculture is based on cloning. In this section, we'll look at how plants and animals have been cloned throughout time—and at the state of the art today.

How Plants Are Cloned

You may not realize it, but if you've worked with plants at all, you've probably done some cloning. The technique is about as simple as it gets. Take a leaf cutting from a plant, help it grow roots, and then replant it. Congratulations! It's your first clone! That new plant has the identical DNA to the original plant from which you took it, and so is a clone of it. This technique is called *vegetative propagation*, and has been used for millennia.

This technique may be second nature to you, and seem rather mundane, but if you stop and think about it for a moment, you'll realize just how amazing it is. The leaf starts off only as a leaf, but it is able to grow entirely new structures—roots and stems—and turn into an entire new plant.

BioDefinition

Vegetative propagation is a kind of asexual reproduction. In asexual reproduction, a mother and a father aren't needed in order to create a new organism—the new organism has the same DNA as the original organism. In sexual reproduction, two sets of DNA are merged, one from the father and one from the mother.

Here's how it works: If you've taken the cutting properly, its end has a group of cells called a callus. These cells in the callus are undifferentiated, that is, they have the ability to turn into many different kinds of cells. The callus grows larger, and then under the right conditions differentiates into roots, and then stems and an entire new plant.

But science marches on, and scientists are never satisfied with something as hit-or-miss as simple of vegetative propagation. So they've come up with a new plant cloning technique, called tissue cell propagation, as shown the following figure.

In tissue cell propagation, cells are taken from root tips, which are the growing part of a root. These cells are then grown in a special, high-nutrient environment, where they become undifferentiated calluses. The calluses are then bathed in plant hormones to stimulate them to grow into entire new plants.

This technique is obviously more complex and more difficult to do than vegetative propagation—I don't know about you, but I don't happen to have a whole lot of spare plant hormones a special nutrient-rich environment in my garden shed. (In fact, I don't even have a garden shed.) Because of that, it isn't used as commonly as vegetative propagation for cloning plants. It's used to grow rare and difficult-to-grow plants and flowers such as orchids.

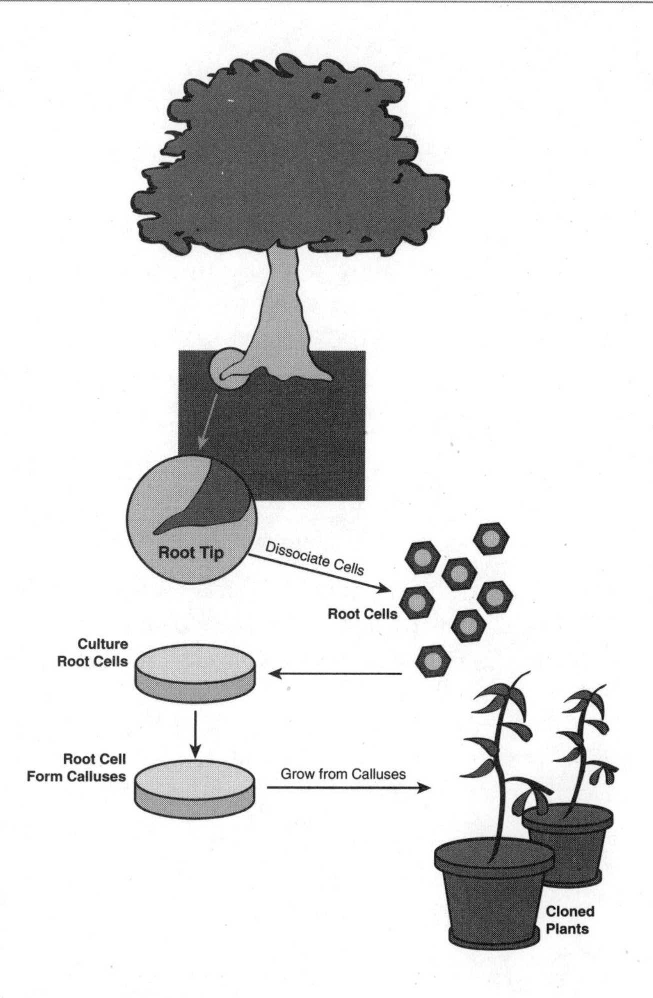

How tissue cell propagation works.

How Animals Are Cloned

When it comes to cloning, plants are easy—animals are hard. After all, we've been cloning plants for millennia, but we've only just gotten around to cloning animals.

> **BioFact**
>
> Spemann did other experiments with salamanders, not all as well-known as the one in which he was able to split the embryo. In one, he constricted the egg with a hair, but didn't completely separate it, and unexpectedly created a kind of Siamese twin salamander tadpole with two heads.

Most people have heard of the cloned sheep Dolly, but cloning didn't start with Dolly. In 1902 the German embryologist Hans Spemann used a strand of hair as a kind of noose and split apart the cells of a two-celled salamander embryo. Each embryo then developed into a full-grown salamander, and each was a clone of the original embryo. In essence, Spemann had induced artificial twinning, mimicking the way that twins are created in nature.

Based on his experiments, Spemann suggested that a "fantastical experiment" be conducted in which an organism would be cloned by transferring nuclear material into an egg. The technology of the time didn't allow that experiment to be done. But in 1952, two scientists, Robert Briggs and Thomas King, were in fact able to do it. They extracted the nucleus from the cells of one frog, and inserted it into an egg cell of another frog whose nucleus had been taken out. The end result: A cloned frog.

In the experiment, the nucleus that was inserted into the egg was taken from an undifferentiated cell from a developing embryo. Undifferentiated cells have the capability to grow into many different kinds of cells. Robert Briggs and Thomas King thought that nuclei from adult cells could never be used to create clones.

Alas, Briggs and King were wrong in their prediction. John Gurdon was able to take the nucleus from an adult skin cell of a frog, and transplant it into a nuclear-less egg from a second frog, as shown in the following figure. The result: A tadpole that is the clone of the frog that donated the nucleus.

Cloning a frog.

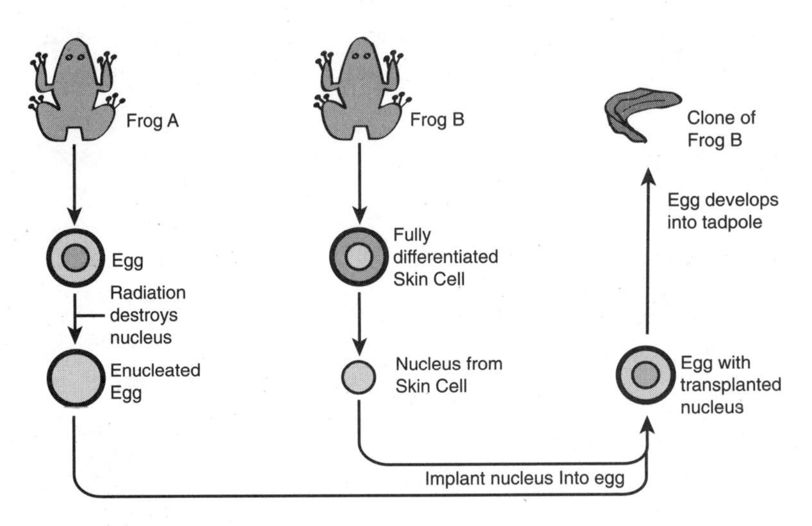

Let's take a closer look at how it works, because the general technique is used commonly in cloning.

First, an egg is extracted from an adult frog (which we'll call frog A), and radiation is used to destroy the frog's nucleus. The resulting egg is called an enucleated egg—one without a nucleus.

Next, a cell is extracted from an adult frog (which we'll call frog B)—in this instance, from a skin cell. The nucleus is extracted from the cell, and is then implanted into the enucleated egg. The egg now has the nucleus of frog B, which contains all of frog B's DNA. The egg becomes an embryo and develops into a tadpole, and because the tadpole has the DNA from frog B, it is a clone from frog B.

BioFact

In Gurdon's experiment, the tadpoles did not grow into adult frogs—they died before that happened. However, the same general techniques he used has been used since on many animals with success.

How Dolly Was Cloned

Gurdon's frog experiments didn't make that big a splash in the world—few people in the general public had heard about them.

But when the sheep named Dolly was cloned, it made front-page headlines all over the world. Not only had a mammal been successfully cloned for the first time, but the implications were clear: If a sheep could be cloned, why not humans?

Dolly was cloned at the Edinburgh-based Roslin Institute by a team led by professor Ian Wilmut through a technique known as *nuclear transfer*. Dolly was cloned not for purely theoretical reasons. Wilmut wanted to genetically engineer sheep and other farm animals so that they could produce proteins in their milk that could help humans—for example, insulin.

BioDefinition

The technique used to clone Dolly is known as **nuclear transfer**, because it transfers the nucleus from one cell into another cell that has had its nucleus removed.

The path was not easy. They attempted to make a clone 276 times and it failed each time. Only on the 277th try did they succeed. Talk about perseverance.

The technique used to clone Dolly builds on Gurdon's experiments with frogs, but takes it several steps beyond. One of the key challenges was to "reprogram" the DNA of the cell that donates the nucleus—the cell that will be cloned. The problem the group faced is that adult cells are programmed to perform certain functions, for example, skin cells are programmed to grow new skin over a wound. They're not programmed to do other things, such as turn into a heart tissue or lung tissue or kidney tissue.

A newly fertilized egg cell contains undifferentiated cells—they can turn into any kind of cell in the human body. So taking the DNA from a differentiated cell of an adult, like a skin cell, and placing it into an egg cell has serious potential problems. The DNA isn't programmed to turn into other kinds of cells, and so the egg cell won't be able to divide successfully, turn into an embryo, and then ultimately turn into an adult animal.

BioFact

Dolly is a cute name for a sheep, but there's a reason behind the name. The cell that donated the DNA was taken from the mammary glands of a sheep. Dolly was named after the well-known country singer and bombshell Dolly Parton, for a pair of obvious reasons.

So you have to "reprogram" the nucleus of that adult cell so that it doesn't just produce skin, for example—it has to be able to produce any kind of skin in the body.

Here's how they solved the problem. As with the experiment with frogs, two cells were required: an adult cell, whose DNA would be used; and an egg cell, which would be the recipient of the DNA.

An egg cell was taken from a Scottish Blackface sheep (which as its name implied, has a black face) shortly after the sheep ovulated. The egg cell was taken then because that was the time that it would be most susceptible to fertilization. An instrument called a pipette was then used to remove the nucleus of the egg cell. The end result was an egg that had not yet been fertilized, and that had no nucleus in it—which means that it had no chromosomes. By itself, it could do nothing.

Then an adult cell was taken from the mammary gland of an udder of a Finn Dorset sheep, that is completely white. The cell was taken from the udder because during pregnancy udder cells grow rapidly. The cell was put in a Petri dish and "starved"—it was denied nutrients that would allow it to grow and divide. Of course, it wasn't completely starved, or else it would die. But it was starved enough to put it into a suspended state. In this way, its state matched the state of the enucleated egg, which was also in a suspended state of sorts because it lacked DNA.

Next, the two cells were placed next to one another, and electricity was used to jolt the cells into merging into a single cell. Once the cells merged into one, another jolt was used to get the new, merged cell to begin dividing. This mimicked the natural burst of energy that accompanies the fertilization of an egg cell by a sperm cell.

The cell now began dividing like any other embryo. After about a week it was implanted into the uterus of a surrogate mother sheep, which was also a Blackface ewe. Why put it into a Blackface ewe, and why have the egg come from a Blackface ewe and the DNA from a white Finn Dorset sheep? So that it would be abundantly clear where the DNA of the resulting clone came from. If the clone turned out to be all-white, its DNA would very obviously be from the Finn Dorset; if it had a black face, it would come from the Blackface ewe. Of course, a DNA test would be able to prove that as well, but DNA tests don't look good on TV. The ultimate result? Dolly, the world's first mammal clone. Its DNA was identical to the DNA of the Finn Dorsett sheep that donated the nucleus.

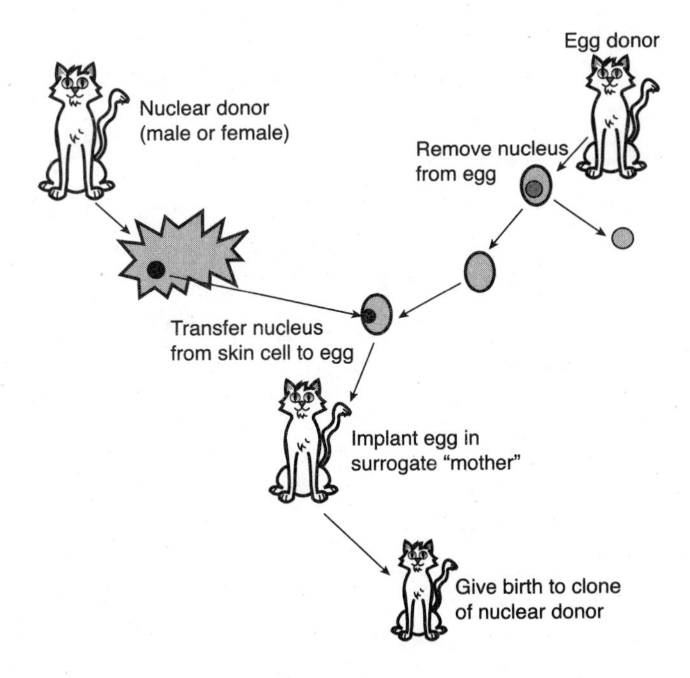

Here's how to clone a cat. Dolly the sheep was cloned in a similar manner.

On Beyond Dolly: More and More Clones

Since Dolly was cloned, scientists have cloned a veritable menagerie of animals. Mice, cattle, pigs, cats, rodents … the list goes on and on. For example, a domestic calico cat was cloned, and given the name Carbon Copy.

One of the more intriguing recent clones was of a racing mule. (Yes, there are such things.) This is unusual because mules are sterile—they don't produce offspring. They are the result of a union of a horse and a donkey. In the process, first a donkey and a horse were bred and the fetus allowed to grow. Researchers then took the DNA from the fetus, and implanted it into the enucleated egg from a horse. The resulting eggs were placed into the womb of a female horse, and the end result was a cloned mule.

BioFact

The cloning of a mule leads to the intriguing possibility that horses could be cloned, and in particular thoroughbred racehorses. Prized thoroughbreds sell for millions of dollars, and so the economic incentive to clone one would be enormous. Not so fast, though. The body that regulates thoroughbred racing has banned cloned horses from racing. In fact, they've also banned any horses that are bred as the result of artificial insemination.

When Is a Clone Not a Clone?

A clone has the exact same genome as the animal or plant from which it is cloned. That doesn't mean, however, that the animal or plant will turn out to be identical to the original. In fact, very often, that is not the case.

Environmental factors affect how organisms turn out to a great degree—as anyone who has followed the nature-or-nurture debate knows. Even human twins raised in the same family end up looking and acting differently as they grow older. One may be heavier than the other, or more athletically talented, or shyer, for example.

The cloned cat Carbon Copy is a perfect example of this. Carbon Copy was a calico cat, as was the cat from which it was cloned. Calico cats have distinctive coat patterns made up of orange, black, and white colors. Yet the calico pattern of Carbon Copy differed from that of the cat from which it was cloned. Although genes made the cat a calico one, other factors decided exactly how that calico pattern looked.

There are other reasons why clones are not actually exact copies. Although cloned DNA always has the same sequence, that DNA doesn't always act the same. There are several reasons for this. One is that the donated cell nucleus already contains some proteins and these proteins influence which genes are actually expressed. So, nuclei donated from two different types of cells from the exact same donor could give somewhat different features to clones even though they contain the very same DNA sequence. Secondly, small chemicals can attach to and imprint the DNA. These chemical attachments differ from cell to cell and can alter how genes are expressed. And the egg contains its own set of proteins that also influence how genes are expressed.

Moreover, the egg does contain some DNA that is not part of the chromosomes. In Chapter 3, we told you that cells contain mitochondria—the cell's power station. Well, mitochondria have their own DNA, and these contain genes that power the cell. So the clone does have a little piece of DNA from its mother. In fact, whether you're a man or a woman, your mitochondrial DNA powerhouse came from your mother.

And Now the Bad News

So far in this chapter, we've talked only about scientific breakthroughs and good news. But there's bad news about cloning as well: It's not very efficient, and generally leads to unhealthy animals that die prematurely. So why is this?

To begin with, it is very difficult to clone an animal. The process has still not been perfected, and it is still very much of a hit-and-miss operation. While the odds tend to be better than Dolly's 277-to-1 odds, they're still not very good. According to a report in the *Los Angeles Times*, barely one percent of cloned embryos survive until birth. Add to that that not all cloning attempts even make it to the embryo stage, and you can see that the numbers are not very good.

But the bad news goes well beyond the numbers. Cloned animals are often very unhealthy animals. Take Dolly, for example. In February 2003, she was euthanized after a series of ailments and only six years of life. Even though Dolly was young for a sheep, she had arthritis, suffered from obesity, and was euthanized after a progressive lung infection.

BioFact

Dolly may have died, but if you'd like to visit her, you still can—and you won't have to go to sheep heaven to see her. Dolly has been stuffed by taxidermists and put on permanent display at Edinburgh's Royal Museum. Those who have seen her report that she's looking as fit—or possibly even fitter—as she did in real life. "She's looking great," a spokeswoman for Edinburgh's Royal Museum told ABC News. "She's on all fours and her head is slightly tilted to one side," the same tilt to her head that she used when greeting human visitors when she was alive.

That didn't come as a great surprise to researchers. They had theorized that she and other clones might be susceptible to premature aging. Dolly's chromosomes were shorter than a normal sheep's because structures in her chromosomes called telomeres were shorter. Telomeres are repeating DNA strings at the ends of chromosomes. Each time a cell divides, the telomere shortens. When the telomeres get very short, the cell has difficulty functioning properly and eventually ages and dies. In that way, many cells are essentially programmed to die after a certain amount of time.

The DNA for cloning is generally taken from cells that have already divided many times, unlike an egg cell, which has yet to divide. So some scientists believe that one of the reasons for the health problems of clones may be that their cells are in essence prematurely aged starting from birth. (For more information about this, see Chapter 4.)

Many other cloned animals face premature deaths and health problems. For example, Japanese researchers cloned mice and compared their life spans with a control group of mice that were the result of natural birth and in vitro fertilization. After 800 days, 83 percent of the cloned mice had died, but only 23 percent of the control group had died. The cloned mice died prematurely from tumors, damaged livers, and pneumonia.

And a group at the Whitehead Institute for Biomedical Research in Cambridge, Massachusetts studied 10,000 genes in the livers and placentas of cloned mice. They discovered that hundreds of the genes had abnormal activity patterns.

Those are just a few examples; many researchers report similar findings.

No one is quite sure why all this happens. In addition to the problem of "old" DNA, there may be other issues as well. The truth is, normal dividing eggs and embryos grow under extraordinarily complex, constantly changing chemical conditions in which gene activity is turned on and off. Embryo division is activated and enhanced by changes in calcium levels, for example, as well as by a chemical called *oscillin* brought into the embryo by the sperm. In cloning, there is no fertilization, and the egg begins to divide in dishes rather than in fallopian tubes and the uterus.

Scientists have even found out that cloned embryos and normal embryos have different nutritional needs. Keith Latham, an associate professor in the Fels Institute for Cancer Research and Molecular Biology at Temple University in Philadelphia, discovered that mouse clone embryos do best when fed glucose, while normal mouse embryos do better with other kinds of sugars.

Many scientists also point out that the problems may be tied to the "reprogramming" that needs to be done in the DNA from adult cells. No one knows what the effects of that reprogramming are.

Where That Leaves Us

Just because there are problems with cloning animals does not mean that animals shouldn't be cloned, say many scientists. There are overriding reasons for cloning animals, particularly when it comes to helping cure diseases, they say. And scientists also expect that they will be able to solve many of the problems in the future as well.

That doesn't mean that the debate about cloning is over—as you'll see in Part 3 it has only just begun.

The Least You Need to Know

- Clones are a copy of another organism that has the exact same DNA as the original.
- Plants can be cloned in a process called vegetative propagation in which a leaf is cut off and a cell mass called the callus develops roots and stems.
- In vegetative tissue cell propagation, root tip cells are grown in a high-nutrient environment, where they become undifferentiated calluses and are then stimulated to grow into new plants.
- When animals are cloned, the nucleus is taken out of an egg and is replaced with the nucleus from a cell from the adult animal to be cloned. The resulting cell and nucleus divides, becomes an embryo, and is implanted so that it can come to full term.
- Many mammals have been cloned, including sheep, mice, cats, rodents, and mules.
- Cloned animals frequently suffer from a variety of debilitating diseases that often affect animals only when they reach an old age.

Chapter 7

How Clones Can Help Humanity

In This Chapter

- Plant cloning in the past
- How plants can naturally clone themselves
- The benefits of plant cloning
- The dangers of plant cloning
- The benefits of animal cloning
- The dangers of animal cloning

Cloning is not about science fiction and offers more than just theoretical benefits. It's a very real new technology that can offer dramatic benefits to humankind in the years to come. Research into plant and animal cloning is relatively new, but has already offered benefits, and as cloning technology matures, it will offer far greater benefits still.

In this chapter, we'll take a look at the benefits of plant and animal cloning, as well as some of the drawbacks. You'll see that when it comes to plants, cloning is nothing new and has been used by humans for millennia. And although animal cloning is newer, it may hold even greater benefits.

How Cloning Has Been Used in Agriculture

If there were newspapers around about 10,000 to 20,000 years ago (give or take a few millennia) when human agriculture had its roots, so to speak, the big news splashed across the front page may well have been "Scientists Clone Plants!" (Of course, there weren't scientists back then, but there weren't newspapers, either, so it all evens out.)

Although cloning continues to make headlines and lead TV news reports, the truth is that plant cloning has been around for as long as human beings have used plants. In fact, it's been around a lot longer than that, because as you'll see a little later in this chapter, plants can naturally clone themselves in a variety of different ways. In fact, plants may have been cloning themselves for about as long as they have been in existence.

Cloning is a natural part of agriculture and plant life, and is frequently used today by backyard gardeners and indoor gardeners as well as by huge, industrial-sized farms. Here are just a few examples:

- When spider plants send off shoots that can then be grown into new spider plants, the resulting new plants are clones of the original.
- Strawberry plants send out runners, and these runners develop into new plants—and those new plants are clones of the original plant.
- When you propagate a plant by using leaf cuttings, you are cloning the original plant through the part of the plant called the *callus*.
- Potato "seeds" used to plant potatoes commercially are in fact not seeds, and are instead potential clones. Each "seed" is a portion of a potato tuber containing a section called an eye. That eye develops into an entire potato plant—and the plant is a clone of the plant from which the section was cut.
- Most of the plants found in nurseries are called cultivars, and have been cloned from other plants via propagation. They are clones because

BioDefinition

A **callus** is a part of a plant, often at the end of a leaf, that contains undifferentiated cells, and can clone an entire new plant. Under the right conditions, the callus differentiates into roots, and then stems and an entire new plant. For more information about plant cloning, see Chapter 6.

that is the most effective way of making sure that each plant has the exact characteristics that gardeners are looking for.

In the past, of course, plant cloning was somewhat of a low-tech affair. Plant breeders would breed plants in the traditional way, to a certain extent by trial and error until they came up with a breed that displayed the characteristics for which they were breeding. Once they had that ideal plant, they could then clone it. For example, the Bartlett pear was created this way at around 1770 in England, and was then cloned. The Delicious apple was bred this way, and then cloned as well. (Each has also subsequently been bred to create new fruit as well.)

BioFact

Although what farmers call potato "seeds" are not seeds, the potato plant can produce true seeds. True potato seeds are used to breed new potato varieties, and are occasionally used to plant potatoes commercially as well.

How Plants Clone Themselves

We've already discussed one way that plants can clone themselves—by sending off shoots or runners. Each of those shoots or runners can then develop into a cloned plant. But there's an odd phenomenon called apomixis in which a plant produces a seed that is genetically identical to itself—in essence, a clone. These plants do not need a mother and a father, and they do not reproduce using pollen. Instead, the seed develops in the plant and carries that plant's identical set of DNA.

More than 300 plant species from 35 different families reproduce this way. It is more common in grasses than other types of plants, although certain asters, daisies, and roses also reproduce this way. Oddly enough, some of the plants reproduce through apomixis and also in the traditional way, by pollen fertilizing an egg. For example, Kentucky bluegrass, often used in lawns, golf courses, and athletic fields, reproduces both ways.

BioFact

If you're looking for a plant that reproduces via apomixis you need look no further than your backyard—and not just if you have Kentucky bluegrass. The dandelion produces seeds this way. So the next time you see dandelion fluff blowing in the wind, ready to invade more yards, you're looking at hundreds of identical seeds waiting to turn into adult clones.

Combining Cloning with Bioengineering

Scientists, always looking to improve on nature, have come up with ways to produce clones more efficiently, and have also combined cloning with bioengineering techniques to create new types of plants. (Note: In this chapter, we are not going to cover bioengineering plants in any detail. For more information, turn to Part 4.)

The most important cloning technique goes by the tongue-busting name of somatic embryogenesis. It is based on the fact that almost all cells of a plant have the ability to become embryogenic—in other words, become like an embryo.

In somatic embryogenesis, cells from plant tissue, often an embryo, are dispersed and cultured in laboratory flasks. The cell's genes are reprogrammed and sometimes new genes are added. Each of these cells can then be tricked into growing into a full plant, and each will be a clone of the original plant. And if new genes were added, these too will be present in the clones.

BioDefinition

A **transgenic** plant or animal is one in which the gene from another plant or animal has been implanted in it.

Scientists can genetically engineer a plant to have a particular characteristic by inserting a particular gene into it, for example, by inserting a gene into soybeans that would give them a particular helpful characteristic, such as being resistant to a certain plant disease. They could then clone the soybeans using this technique and grow it that way.

This cloning technique can be used in a variety of ways to help agriculture. For example, scientists at Penn State University are helping to improve the cultivation of cocoa plants by cloning the most productive types of cocoa plants. Ultimately, those clones will be planted in cocoa plantations. It can be very difficult to produce only the most highly productive cocoa plants using normal breeding methods.

Scientists are also using bioengineering to maintain high-quality hybrid food crops. Hybrids can be bred naturally to have particularly desirable traits, for example, to taste better, last longer, have more nutritional value, and so on. But there is a problem with hybrids—because they reproduce sexually by combining with less-desirable plants, those good traits can disappear or change over the generations. They can disappear or change because sexual reproduction means that each generation gets a mix of genes from its parents, and so the traits may change over time.

With apomixis, on the other hand, the offspring will be exact clones of the original plants, and so they will maintain the same traits. Scientists are investigating ways in which hybrid food crops could be bioengineered so that they would reproduce using apomixis. That way, each succeeding generation will be identical to the previous one.

Corn has already been bioengineered this way. A portion of the chromosome that directs apomixis was implanted into corn from corn's distant cousin, a plant called eastern gamagrass. The resulting corn/gamagrass hybrid is able to reproduce via apomixis. More work needs to be done before the plant can be used commercially, but it is a first step.

The Benefits and Dangers of Plant Cloning

Human cloning and animal cloning make the news and incite controversy; plant cloning tends to draw a big ho-hum, at least in the United States. But though they're not as well known, cloning plants offers a great many benefits and has some potential dangers as well.

Note that the benefits and dangers of plant cloning are very closely related to those offered by bioengineering. So this chapter will only cover those that are most directly related to cloning. For a more detailed discussion of the benefits and dangers of bioengineering plants, turn to Chapters 19, 23, and 24.

> **BioFact**
>
> If you're looking to find out more information than you perhaps ever wanted to know about agriculture and bioengineering, go to Ag Biotech InfoNet at www.biotech-info.net/. It is sponsored by a consortium of scientific, environmental, and consumer organizations.

Why Plant Cloning Is Good

Those who favor scientific cloning of plants say that it has many benefits for consumers as well as the environment. Here are some of the primary reasons they say that plant cloning should continue:

- **It can increase food production and create healthier foods.** Cloning high-quality hybrids means that the best plants for agriculture can be grown, and will not die out over time. It also can help feed the poor in Third World nations, by making available the high-quality hybrids, and making it easy to distribute those hybrids. And it can make available resistant crops that are now ravaged by insects, disease, or poor soil conditions.

> **BioFact**
>
> If you have kids who want to try plant cloning at home—or if you want to try it yourself—go to www.sciencekit.com and search for "plant cloning." They have kits that will help you clone common plants, and also have complete kits for cloning African violet plants.

- **It can help local economies.** Many local economies are built on agriculture. So if cloning can help local agriculture, it will help local economies. This is particularly important in Third World countries. For example, the Penn State project that clones cocoa plants will help keep economies in cocoa regions healthy by ensuring that only the healthiest cocoa plants are grown.
- **It can help preserve endangered species**. By cloning plants that are endangered, those species can be preserved so that they don't die out, and their genetic heritage can be retained.

The Dangers of Plant Cloning

Those who warn about the dangers of plant cloning don't want to try and ban cloning, because it has been used in agriculture for thousands of years. Rather, they focus on the dangers posed when cloning is combined with bioengineering. They warn about these cloning dangers:

- **Cloning cuts down on biodiversity.** When plants are cloned, the offspring are identical to the original plants. That means that there is a great deal less biodiversity than when plants reproduce naturally.
- **Cloning could decrease resistance to disease and insects.** If only a very few food crops are used, and those are clones, then a single disease or insect could wipe out enormous amounts of agriculture and potentially lead to famine. If there were more genetic diversity in the crops, they would not be subject to the same disease.
- **Clones put economic power in the hands of large corporations, not individuals.** If farmers cannot grow crops from seeds because those crops can only be grown from clones, it means that they are dependent on corporations to provide them with plants they can clone. That puts them at the mercy of large corporations.
- **Clones could have unexpected effects on our health.** Some people worry that bioengineered plant foods could include harmful substances as byproducts of cloning. None of these has so far been found.

The Benefits of Cloning Animals

Scientists clone animals not for theoretical purposes—they have real-world, practical ends in mind. Often, but not always, to get to that practical end requires the use of bioengineering techniques, not just cloning.

For example, scientists hope that some day pigs can be cloned and their organs harvested for human use. But in order for that to happen, bioengineering must be used so that certain genes of the pig are "knocked out" so that the organs will not be rejected by the human body in which they are implanted. After the genes have been knocked out, clones will be made of pigs, and their organs used for human implantation. That's the theory, anyway.

BioWarning

Cloned animals are often born larger than noncloned animals—and sometimes dramatically so. For example, when an endangered animal called the banteng (a kind of cattle in Bali) was cloned to help ensure that it did not become extinct, one of the two cloned animals was born twice the normal size—80 pounds instead of 40 pounds. It was euthanized.

Cloning animal organs and implanting them in humans is many years away in practice, if it in fact can ever happen. But that's just one example of how cloning is generally used in concert with bioengineering. In this section of the chapter, we'll focus on cloning; for more information about bioengineering, see Part 4.

Although there are many potential uses for cloned animals, currently these are among the uses to which clones might be eventually put:

- **To create organs for transplanting into humans.** As we earlier explained, scientists hope that one day pig organs may be able to be successfully transplanted into humans from genetically modified pigs. In fact, transgenic pigs with some genes knocked out of them have already been cloned, first at the University of Missouri in October of 2001. More than 62,000 people in the United States are on waiting lists to receive donated organs, but the demand far outstrips the supply, so not all people receive them. Mass-producing organs for transplantation can ensure that anyone who needs an organ can get one.

BioFact

Advanced Cell Technology, based in Massachusetts, says it was able to grow functional kidney-like organs from cloned stem cells and transplant them into cows—and the cows did not reject the tissue because the stem cells had the cow's identical DNA makeup. The organs, when transplanted into the cows, essentially functioned as kidneys, and were able to excrete metabolic waste products through a fluid like urine.

- **To mass-produce pharmaceuticals.** A gene can be inserted into an animal to make that animal produce a particular protein in its milk—for example, insulin.

If that animal were then cloned, pharmaceuticals could be mass-produced at low cost. Many pharmaceutical firms are funding research into this area. In fact, that was the primary driving force behind creating the most famous clone of all, Dolly.

- **To create animals that can be used in laboratories.** Scientists frequently use animals such as mice to experiment on—and very often, drugs and certain medical procedures need to be first tested on animals before they can be tested on humans. Animals particularly well suited for lab experiments can be bioengineered and then cloned. For example, mice can be bioengineered and given a gene that will cause them to get cancer, and those mice can be given anti-cancer drugs to see whether the drugs work. Once the mice are bioengineered, they can be mass-produced by cloning.

- **To create animals more suited for farming.** Animals can be bioengineered to be better suited to farming, and then mass-cloned. The University of Vermont and the U.S. Department of Agriculture, for example, bioengineered a cow by giving it a gene that produces a protein called lysostaphin that can protect the cow against mastitis disease. Mastitis stops a cow from producing milk by destroying the cow's milk-secreting cells in its udder. The cow was then cloned—and the hopes are that it will prove to be resistant to mastitis so that it can be mass-produced.

BioFact

The federal National Institute of Science and Technology has given $4.7 million to help fund research into "cloning" chickens on a massive scale for potential use in chicken farming. The research will go toward developing disease-resistant chickens that grow faster than current chickens and use less food, and then using a cloning-like method on them. Birds can't truly be cloned because bird eggs can't be removed and implanted, as can eggs from other animals, such as mammals. But farmers can use a cloning-like technique to inject donor cells into eggs, producing a "chimera" in which 95 percent or more of the chicken will develop from the donor cells.

- **To save endangered species.** Animals endangered in the wild can be cloned and kept safely in zoos, or else re-introduced into the wild. This has already been done. An endangered cattle-like animal called the banteng was cloned from a cell that had been frozen for safekeeping in the San Diego Zoo's "frozen zoo," which keeps cell and DNA samples of many endangered species. The banteng's skin cell was combined with the egg of a cow whose nucleus had been removed, and so the egg had the DNA makeup of a banteng. The resulting cloned embryo was then implanted into a cow, and the cow gave birth to a banteng.

- **To clone pets.** Some pet owners can't seem to let go of their favorite pets—and so they may want to clone their pets. In fact, companies have already sprung up, such as Genetic Savings and Clone (GSC) at www.savingsandclone.com, that for a fee will provide a "gene bank" for your pet so that you can clone it after your pet dies, and when cloning technology becomes available. In fact, GSC is funding research at Texas A&M University into cloning pets, which led to the cloning of the first cat, called Copy Cat.

BioWarning

People who clone their pets because they want another just like the one they love may be in for a rude surprise—the cloned animal will in all likelihood be very different from the original. A study of cloned pigs done by North Carolina State University's College of Veterinary Medicine found that cloned pigs looked different from one another, and also had different temperaments and preferences for food. So even though they had the same genetic makeup, they looked and acted differently. In fact, the cloned cat Copy Cat does not look exactly like the cat from which it was cloned, and has a different temperament as well. These differences probably arise from how the growing embryo is influenced by environmental factors in the mother and the egg.

The Dangers of Animal Cloning

As you can see, there are many potential benefits to animal cloning. But that doesn't mean that the technique is risk-free. It can in fact pose dangers as well, such as these:

- **It may lead to the creation of new diseases.** Many of the newest, emerging diseases, such as AIDS and Severe Acute Respiratory Syndrome (SARS) are caused when viruses jump the species barrier—in other words, when a virus originally from an animal makes the leap to man. Some people worry that if organs from cloned animals were transplanted into humans, it could lead to the creation of deadly new diseases.

- **It can cut down on biodiversity.** If agricultural animals are mass-cloned, there will be a much smaller animal gene pool. The mass-produced animals could be subject to the same disease, while if there were more genetic diversity in the crops, they would not be subject to the same disease.

BioSource

Animal cloning is controversial in some quarters, particularly among animal-rights activists. For more details about the debate about animal rights, cloning and bioengineering, turn to Chapter 20.

- **It can be cruel to animals.** As a general rule, cloned animals are not healthy animals. Therefore, creating a new cloned animal may subject it to a short, dismal life.
- **It can be a waste of valuable research resources.** No society has an endless amount of money and research resources, and so society must determine where to spend its time and research dollars. Some people contend that spending it on cloning, and particularly on things such as cloning pets, is a waste of resources, which could be put to better use elsewhere.

The Least You Need to Know

- Humankind has been cloning plants for agricultural purposes for thousands of years, and some of the same techniques are still being used today.
- Some plants naturally clone themselves through apomixis in which the plant produces a seed that is genetically identical to itself—in essence, its own clone.
- Bioengineering techniques, combined with plant cloning, can offer many benefits such as the development of higher-yielding or disease-resistant crops.
- Among the dangers of plant cloning are the loss of biodiversity, which can lead to entire crops being wiped out by a single disease or pest.
- Bioengineering techniques, combined with animal cloning, may yield many benefits, such as the mass-production of pharmaceuticals, the creation of organs for human transplantation, and more efficient farm animals.
- Among the dangers of animal cloning are a loss of biodiversity and the possibility that cloned organs for human transplantation may carry new diseases into humans from animals.

Chapter 8

The Promise of Therapeutic Stem Cell Cloning

In This Chapter

- Understanding stem cells
- The difference between embryonic stem cells and adult stem cells
- How therapeutic stem cell cloning works
- How stem cells may be able to cure a wide variety of diseases and conditions
- What the federal government says about stem cell research

Of the many therapies related to biotechnology and cloning, the one that has received perhaps the most publicity, and may hold out the most promise, is called by the general term therapeutic stem cell cloning. The term is in fact something of a misnomer, as we'll see in this chapter, because it doesn't involve cloning in the way that you normally think of it.

In fact, most of the research and potential therapy has to do with stem cells rather than cloning. In this chapter, you'll learn what stem cells are, how they can be used, and you'll get an understanding of therapeutic stem cell cloning as well.

What Are Stem Cells?

Your body is made up of many different kinds of cells—heart cells, blood cells, skin cells, liver cells, muscle cells … the list, as you might imagine, is very long. Each of those cells is specialized and devoted to a specific function.

But those specialized cells didn't arise out of nowhere—you did not wake up one day a whole human being, with all those cells in you. (At least, unless you were Frankenstein, you didn't wake up that way.) All of the cells in your body, literally from the tip of your head to the bottom of your feet, began life as a single cell—an egg cell. After the egg cell was fertilized by a sperm cell, the cell quickly began to divide and divide again. Ultimately the resulting cells began to differentiate, until you became you, for better or worse (for better, we hope).

BioFact

The umbilical cord contains a high concentration of stem cells. Because of that, some people have used medical services to save the umbilical cords of their newborns, in case those stem cells are ever needed for therapeutic reasons later in life.

That initial egg cell, and the first groups of cells that began dividing were still undifferentiated. They weren't yet heart cells or skin cells or liver cells or any other kind of specialized cell because at that point, you had no heart, no skin and no liver—you had no organs of any kind. You were just a tiny clump of fast-dividing cells.

These cells each had the ability to turn into any kind of cell that your body would ultimately need, whether it be a skin cell, heart cell or any other kind of cell.

They were stem cells—one kind of stem cell, as you'll see in this chapter, because there are other types as well. Stem cells have the ability to turn into differentiated cells. They stay as stem cells until they receive a signal to differentiate into another kind of cell. In addition to being able to differentiate into other kinds of cells, stem cells also divide very rapidly.

Both of these features mean that stem cells are ideally suited for a wide range of potential medical uses. Because they can develop into any other cell in the body, they could theoretically be used to replace damaged or diseased tissues and cure a wide variety of diseases, potentially even become a miracle cure of sorts. Because they divide so rapidly, they could do so quickly. As you'll see later in this chapter, they have many other potential uses as well, for example, possibly even cloning organs.

The Different Kinds of Stem Cells

There are two general different kinds of stem cells, stem cells from an embryo, and stem cells from an adult. While they are related, there are significant differences between the two, and they hold out different potential to cure disease, with embryonic stem cells holding out much more promise.

Stem Cells from Embryos

Embryonic stem cells can be obtained in a number of ways, most notably from embryos that would otherwise be discarded in fertility clinics. There is, in fact, more than one type of stem cell that can be harvested from embryos:

- **Pluripotent stem cell** This is a general term that refers to a stem cell that can differentiate into the types of cells that can be formed from the three "germ layers" (*mesoderm*, *endoderm*, and *ectoderm*) of the embryo. As a practical matter, this means that *pluripotent stem cells* can develop into any cell in the body. Pluripotent stem cells can only come from human embryos in the very earliest stages of development and from fetal tissue that would ultimately develop into the gonads, the male or female reproductive organs.

BioDefinition

Pluripotent stem cells are stem cells that have the ability to turn into any other cell in the body. Stem cells from embryos and fetal tissues have this ability, while stem cells from adults can only turn into certain kinds of cells.

The **mesoderm** is the middle germ layer of an embryo. It develops into tissues and structures including muscle, bone, skin and tissue. The **endoderm** is the innermost germ layer of the embryo and develops into portions of the digestive tract. The **ectoderm** is the outermost germ layer and develops into the nervous system, the epidermis (outermost layer of the skin), and the lining of body cavities such as the mouth.

- **Embryonic stem cell** Embryonic stem cells come from an embryo that is four to five days old and is called the blastocyst. Cells from what is called the inner cell mass can be removed and cultured as embryonic stem cells. Embryonic stem cells are pluripotent and can develop into any cell in the body.

- **Embryonic germ cell** Embryonic germ cells come from fetal tissue—specifically, from the area of the embryo that ultimately develops into the testes or ovaries. Embryonic germ cells are pluripotent and can develop into any cell in the body.

Stem Cells from Adults

Stem cells in adults are found in some adult tissues, such as the bone marrow, brain, blood, cornea, and retina of the eye, liver, skin, and muscles among other tissues. They create replacement specialized cells for the tissues in which they are found—so if cells in those tissues are damaged, diseased, or lost by normal wear and tear, stem cells can generate new specialized cells to replace them. Unlike embryonic stem cells, they cannot normally differentiate into any cell of the body. They can only differentiate into the type of tissue where they are found. So stem cells in the cornea, for example, cannot normally develop into nerve cells, and stem cells in the pancreas cannot normally develop into eye cells.

However, in the last several years, experiments have shown that under certain conditions, some adult stem cells can differentiate into tissues other than the ones in which they are found. The capability to do this is called plasticity. Experiments have shown that in the laboratory blood stem cells can be tricked into becoming neurons, and liver stem cells can produce insulin, for example. How this happens is not well understood, and it is unclear how plastic adult stem cells can ultimately be.

BioDefinition

Adult stem cells are considered unipotent—under normal conditions they can only give rise to one type of differentiated cell.

Adult stem cells are rare, and difficult to identify and grow in the laboratory. Unlike embryonic stem cells, they do not replicate themselves for very long periods when grown in culture in the laboratory.

How Stem Cells Work

Because stem cells can differentiate into any cell in the body, they hold out enormous promise for curing a wide variety of disease, as we'll see later in this chapter. But before we look at that, we'll first examine how scientists work with stem cells, and how stem cells differentiate into different cells in the body.

Working with Embryonic Stem Cells

A technique known as cell culture grows cells in the laboratory, and it has been adapted to grow embryonic stem cells. Human embryonic stem cells are transferred to a culture dish that contains nutrients that the cells use in order to survive. The embryonic cells grow in the dish, and after several days begin to proliferate enough so that they began to crowd the dish. Cells are then removed and dispersed to seed new culture dishes, where they grow, and when those dishes become full, the cells are put into even more dishes. In this way, a large number of embryonic stem cells can be grown from just a few cells. After six months, for example, the original 30 embryonic stem cells that were taken from the inner cell mass of an embryo can grow into millions of cells.

Care needs to be taken when growing embryonic stem cells in this way, because they are essentially genetically programmed to differentiate into specialized cells. If the proper care isn't taken, they will differentiate on their own. For example, if the cells are allowed to clump together to form what are called embryoid bodies, they will spontaneously form muscle cells, nerve cells and other kinds of cells.

When stem cells are taken from an embryo and cultured in the laboratory this way, they are referred to as a stem cell line. Scientists carefully document details about each line. Because every embryo is genetically different, each stem cell line will be genetically different as well. Databases of information are published about what stem cell lines are available, and contain a great deal of data about each line. But they all have in common that they can differentiate into cells of many types of tissues.

BioFact

It has taken scientists decades to perfect ways of growing stem cells, ensuring that they were in fact stem cells and not some other kind of cell, and then culturing them in the laboratory. In 1981, scientists were able for the first time to grow mouse embryonic stem cells in the laboratory. It took scientists almost 20 more years to be able to do the same thing for human embryonic stem cells. In 1998, James Thomson at the University of Wisconsin-Madison developed and grew the first human embryonic stem cell line.

Embryonic stem cells will continue to divide indefinitely and stay as stem cells under the right conditions. The key to using them in therapy, however, is to get them to differentiate into different kinds of cells—for example, into brain cells that can be used to replace or repair damaged brain tissue.

Scientist still don't understand how to do this, although they have developed a number of techniques (called protocols or "recipes") for doing so. The cells are allowed to clump into embryoid bodies, and then are treated in a variety of ways to force them to grow into specific cells—for example, insulin-secreting pancreatic cells, or neurons (nerve cells) that secrete the chemicals dopamine and serotonin.

What Is Therapeutic Cloning?

The phrase "therapeutic cloning" is frequently used, and almost as frequently misunderstood. It in fact has nothing to do with cloning human beings or animals, and instead refers to a kind of research and therapy using stem cells.

The phrase is used differently by different people. In the most general sense, it refers to the general technique of gathering stem cells from an embryo, and inducing those cells to differentiate into specialized cells, with the hope of using those differentiated cells to cure or alleviate a disease or condition, for example to regenerate diseased heart tissue.

This shouldn't be controversial but it is because of where stem cells come from—the human embryo. For example, one way to get stem cells begins by removing a nucleus from a human egg and replacing it with the nucleus and DNA from a cell of a person who is a DNA donor. That egg, with the donated DNA in it, is induced to start dividing. Stem cells are harvested from the resulting embryo. Those stem cells could then theoretically be used to create tissue or cells of some kind that can help cure a disease from which the cell was taken. If the tissue or cells are transplanted into the donor, his body won't reject them, because they contain his genetic makeup.

Experiments have already been undertaken to prove that this might work. It was first attempted in November 2001 by a biotechnology company in Massachusetts called Advanced Cell Technologies (ACT). ACT collected eggs from women's ovaries, and then removed the DNA from the eggs. A skin cell was inserted into the egg as a way of giving the egg a complete set of DNA (an egg has only one half the number of human chromosomes, and gets the other half from sperm). The egg was then stimulated with a chemical called ionomycin and the egg began to divide. The experiment was only partially successful: Of eight eggs, only three began to divide, and only one of those reached a stage where it had three cells. Still, it proved that therapeutic cloning is theoretically possible.

Working with Adult Stem Cells

Adult stem cells are more difficult to work with than embryonic stem cells. They are rare, difficult to isolate, purify, and grow. Unlike embryonic stem cells, adult stem cells do not clump together and differentiate on their own in culture dishes. Tests have shown that when mouse adult stem cells are removed from the culture dish in which they are being grown, and injected into a mouse with a compromised immune system, tumors can develop. Unlike embryonic stem cells, they do not proliferate to a great degree in culture dishes, and often do not proliferate at all. So it is much harder to get large quantities of these cells. Recently, scientists have found ways to grow adult mouse stem cells efficiently in the laboratory and this may lead to treatments for leukemia and other blood diseases.

As you can see, embryonic stem cells are much better suited to study and for treating disease than are adult stem cells. It is difficult to obtain large quantities of adult stem cells, they can be difficult to grow, they cannot differentiate into any cell in the body, and they don't live as long in culture as do embryonic stem cells. However, they do have one potential benefit. Because they are taken from an adult, if they are taken out of that adult's body and then later transplanted into the body, they will be less likely to be rejected as foreign matter.

BioFact

Adult stem cells have been used in gene therapy trials as a way of transferring healthy genes into the body. Stem cells—frequently hematopoietic, or blood-forming stem cells—are taken out of the body, and mixed with a harmless virus that will deliver the healthy gene into cells. Then the stem cells, now with the healthy genes, are inserted back into the body, where the healthy gene can do its work. Stem cells are used in this way for a number of reasons, especially because they are self-renewing, and so will continue to produce cells with the healthy gene, which may reduce or eliminate the need for continually administering the therapy. (For more information about gene therapy, see Chapter 22.)

The Benefits of Stem Cells

Study of the therapeutic uses to which stem cells can be put are relatively new, but many studies show a wide variety of potential benefits. And it is not only the stem cells themselves that may provide cures for disease—studying how stem cells differentiate into specialized cells may provide clues in how to fight disease as well.

Turning genes on and off is central to the process of turning stem cells into specialized cells. Many serious diseases and conditions, such as cancer and birth defects, are caused when cells divide and differentiate abnormally, so the study of stem cells may lead to therapies to cure those diseases and conditions.

BioFact

A technique may enable scientists to create embryonic stem cells from an adult, and use the cells therapeutically. Scientists at the University of Pennsylvania have turned embryonic mouse cells into an unfertilized egg. If that could be done with humans, it would become a major source of human eggs. With many eggs available, the nucleus from an adult human cell could be put into an egg that had its nucleus removed, and the egg could then be induced to turn into an embryo. Embryonic stem cells could be taken from that embryo and turned into tissue that could cure disease in the adult—and because the cells would have the same genetic makeup as the adult, the tissue would not be prone to being rejected by the body.

There are many ways that stem cells can potentially be used directly to cure diseases as well. Here are several ways in which stem cells may be used:

- **Brain cell transplantation** Stem cells could be differentiated into brain cells and nerve tissue and be transplanted into the body. This could help cure diseases and conditions such as strokes, spinal cord injuries, and degenerative brain and nerve conditions such as Parkinson's Disease. Studies have already been done into transplanting fetal tissue into those who have Parkinson's disease.

BioFact

Stem cells have actually been used therapeutically for quite some time. For example, bone marrow transplants are used to repopulate the bone marrow and replenish blood cells after chemotherapy or radio-therapy. In a bone marrow transplant, stem cells from the bone marrow of a matched donor are transplanted to restore blood function.

- **Treating diabetes** Insulin is produced by structures in the pancreas called the islets of Langerhans, and diabetes is caused when not enough insulin is produced by them. Scientists have been able to create insulin-creating cells from mouse stem cells—and those cells assemble themselves into structures that closely resemble the pancreatic islets. Potentially, stem cells could be used to cure diabetes in this way.
- **Replacing skin** Stem cells could be used to create skin that can be used for skin grafts when treating burn victims.
- **Cloning organs** In theory, entire organs could be cloned and transplanted, and because the DNA of the organ would be the same as the person from which it was cloned, it would not be likely to be rejected. DNA would be extracted from the person who needed the transplant, it would be implanted into an *enucleated egg*, and from the resulting embryo stem cells would be harvested. Those stem cells could then generate an organ. This, of course, is only theoretical at this point, and scientists have not come close to doing something like this.

BioDefinition

An **enucleated egg** is one in which the nucleus has been removed.

- **Curing muscular dystrophy** Scientists at the Children's Hospital of Pittsburgh and the University of Bonn discovered adult stem cells in muscles that can differentiate into a variety of cell types. They hope that the cells may one day be able to treat muscle diseases such as muscular dystrophy.

There are many other kinds of diseases that stem cells may be able to cure. And these kinds of therapies are not merely theoretical at this point—some are already being put into practice. For example, stem cell therapy may have saved the life of 16-year-old Dimitri Bonnville. He was accidentally shot in the heart with a nail gun, which led to a massive heart attack. In order to live, doctors said, he would have needed a heart transplant, but none was available. So instead, they tried an experimental stem cell therapy. He was given drugs to stimulate the production of stem cells in his blood. Those stem cells were harvested, then infused into the artery that supplies blood to the heart. The cells apparently grew into heart tissue, because the teen survived, and his heart began functioning more normally.

> **BioSource**
>
> A great site for explaining the basics of stem cell and stem cell research is run by the Genetic Learning Center at the University of Utah. It includes entertaining animated explanations. Find it at gslc.genetics.utah.edu/units/stemcells/.

Not all stories have a completely happy ending, though. Although the teen was helped by the therapy, the Food and Drug Administration said that animal studies need to be done before others can use the therapy.

The Feds Step In

With the astonishing number of benefits that may be offered by stem cell research and therapeutic cloning, one wouldn't expect that it would be at all controversial. But in fact, it has become one of the major flashpoints in what some people refer to as the "cultural war."

We'll explore the issues more fully in Chapter 14. In a nutshell, here it is—and the upshot of what eventually happened.

Anti-abortion activists led a fight against the use of embryonic stem cells for any purpose at all, claiming that a fertilized egg is a human being, even if it exists only in a laboratory dish. Supporters of the use of embryonic cells countered that laboratory embryos are not human beings, and that, at the least, only embryos normally discarded during fertility treatments would be used.

> **BioFact**
>
> Many of those who oppose abortion support the study of embryonic stem cells, and disagreed with President Bush's ban on federal support of much embryonic stem cell research. For example, prominent Senatorial abortion foe Orrin Hatch (R-UT), wanted President Bush not to restrict the use of federal funds for embryonic stem cell research.

President George W. Bush sided largely with the anti-abortion activists, and on August 2001 banned the use of any federal funds for research on embryonic stem cells, except for on certain embryonic stem cell lines that had been created before the date of his announcement.

Scientists warn that the restriction will slow down research, and possibly cost many lives. They also warn that because other countries have no such ban on research, the United States will fall behind in research in this area.

The Least You Need to Know

- Embryonic stem cells are found in the early stages of an embryo. They have the ability to turn into any kind of cell that your body would ultimately need, whether it be a skin cell, heart cell, or any other kind of cell.
- Adult stem cells are found in several tissues in the body. Under normal conditions, they can only develop into cells related to the tissue in which they are found.
- Embryonic stem cells are better suited than adult stem cells for research and therapy because they can more easily be isolated and grown and can differentiate into a wider variety of tissues.
- Scientists are investigating the use of stem cells to cure a wide variety of diseases—for example, using them to cure brain diseases such as Parkinson's Disease by having them grow into brain tissue.
- President George W. Bush banned the use of federal funds for research involving embryonic stem cells, unless the stem line cells were created before August 9, 2001.

Chapter 9

The Infinite Twin: Cloning Human Beings

In This Chapter

- A look at the rogue's gallery of wanna-be cloners
- Can human beings be cloned?
- Why cloning humans and other primates is not yet possible
- The reasons that some people give for cloning humans
- The medical dangers associated with human cloning

Cloning sheep, pigs and mules may fascinate the public, but that's largely not because we like to see mirror images of animals. It's because of the implications that such cloning has for human beings. If high-order mammals can be cloned, people wonder, are human beings next?

In this chapter, we'll take a look at the scientific aspects of human cloning. We'll start off by surveying those people who claim that they have already produced clones or will soon produce clones, then look at the science of human cloning, and see why some scientists say it may never be possible, Dolly the sheep notwithstanding. Then we'll end the chapter with a brief look at why some people favor cloning, and the scientific dangers of cloning. We'll cover both of these issues in much more detail in Part 3.

Who Claims to Be Cloning?

We live in a mediacentric world whose patron saint appears to be P.T. Barnum. So it should come as no surprise that TV, newspapers, and magazines have devoted sizable amounts of time and space to the often-dubious claims of those who claim to have cloned human beings, or say they will soon clone human beings. After all, if you can't display a circus freak, Tom Thumb, or a false "Fiji mermaid," why not draw a crowd by saying you've cloned a human?

As we'll see later in this chapter, science has not yet advanced to the point where human beings can yet be cloned, and there are those who say that it may never be possible to clone them at all. That hasn't stopped people from saying that they've already accomplished the task, however, or that they will soon accomplish it.

No book about cloning could be complete with at least a partial rundown of those who claim to have cloned human beings. So here then is a gallery of sorts of those what have claimed to clone, and what has happened to them since their claims.

The Raelians: Cloning's Space Shots

Cloning media madness reached its peak in December of 2002 when a cult called the Raelians claimed to have cloned the first human being. Though they offered no evidence to back their claim, news organizations from around the world trumpeted the cult's assertions. The Raelians said that through their company, the U.S. based Clonaid, they had cloned a baby girl named Eve, who was born to a 31-year-old woman in the United States.

That the media took the cult's claims seriously and reported on them at all was fairly astounding. This was a cult, after all, that was founded by an ex-racing car journalist in 1973 and who believes that humans were created by space aliens who had mastered bioengineering and had cloned human beings. This was all revealed to cult leader Claude Vorilhon (who took the name of Rael), when a space alien landed his UFO near him and spoke to him in unimpeachable French, he says. (Perhaps if the alien spoke with an America accent, Rael wouldn't have been taken in so easily.) The alien, he said, was about four feet high, had olive skin, long dark hair, and almond-shaped eyes.

BioFact

The Raelians' original symbol was a Swatstika inside a Star of David, but they changed it in 1990, because they wanted to improve their relations with Israel. Why did they care about their relations with Israel? They wanted to convince the Israeli government to let them build an embassy in Jerusalem for the space aliens, which the Raelians call Elohim. Needless to say, the embassy has never been built.

Since its founding, the cult has grown, although it's unclear how large it truly is. It claims to have 60,000 members in 84 countries, but there is no way to verify that claim (just as there is no way to verify its claim about cloning). But on a membership drive in the State of Washington several months after its claim of having cloned a human being, it was discovered that there was only a single cult member in the entire state.

Since its initial claim, the Raelians say they have cloned another baby in South Korea, although they provided no evidence of that claim either.

The cult appears to have tried to cash in on its moment in the media sun. *Wired Magazine* reports that Rael gets $100,000 per speech, and the *Las Vegas Review-Journal* reports that Clonaid is selling its "cloning machines" for $9,000 per machine. The Clonaid website (www.clonaid.com) doesn't sell the cloning machines, but it does sell a variety of other services, including a service called "Insuraclone" for $200 per year, in which a Clonaid employee will collect your cell samples and store them in a secure place. It is also selling "genetic repair kits" for an unnamed price, as well as a "Clonapet" service, also for an unnamed price, in which Clonaid will clone your pet.

Severino Antinori and the Italian Connection

The mass media may have discovered cloning when it reported on the Raelians, but the cult was not the first group or individual to say that it was pursuing cloning. In 1998 the Italian doctor Severino Antinori announced that he would embark on cloning human beings.

BioFact

The year 1998 seems to have been a big year for would-be cloners, because in that same year, someone with the too-fitting name of Dr. Richard Seed—who is neither a medical doctor nor does he have a background in medicine; his Ph.D. is for physics—announced that he was setting up a clinic to clone human beings. Seed was never able to raise the money to start a clinic in the United States and so he said he would set one up in Japan. Ultimately, none of it worked out and Seed has faded from view.

Dr. Antinori, unlike the Raelians, at least had a medical background, and had done work related to the field. He ran a fertility clinic in Rome, and was best known for his work in helping women in their 50s and 60s give birth using *in vitrio* fertilization. He first made the news when he helped a 63-year-old woman give birth to a child, the oldest known woman to give birth.

BioDefinition

In vitrio fertilization is a technique in which eggs are taken from a woman, fertilized outside her body, and then the resulting embryo is implanted in her uterus several days later, so that the woman can give birth.

In 2001, he made news again when he said that he was closer to cloning a human being, and would clone one in an unnamed Mediterranean country. Then in 2002 he claimed that he had implanted a clone into a woman, who would give birth in January 2003. However, news of the event was never delivered, and Dr. Antinori has largely dropped out of the news since then.

Panos Zavos and Cloning

At one of Dr. Antinori's press conferences, he was teamed with the fertility researcher Dr. Panos Zavos. Together, they said, they would clone a human being.

But rather than cloning, the doctors were splitting. In May, 2002, Dr. Zavos dissociated himself from Dr. Antinori and said that he was assembling his own team to create a human clone.

Dr. Zavos has not made any claims that he has yet cloned a human, and has criticized claims like those of the Raelians that a human has already been cloned. He has, however, claimed that he has been able to grow a human cloned embryo to the eight-cell stage, an important first step toward cloning humans. He says that he will test cells from the embryo to see if they contain genetic abnormalities, and if they don't, he will implant it in a surrogate mother.

A number of scientists, however, note that technical details were lacking in Dr. Zavos's announcement, and are not convinced that he has done what he has said.

Can Human Beings Really Be Cloned?

With all the debate, claims, and counter-claims about human cloning, a question seems to have been inconveniently forgotten—whether human beings can, in fact, be cloned.

After all, even though sheep have been cloned, cows have been cloned, cats have been cloned, mules have been cloned, and other animals have been cloned, people are not sheep, cows, cats, and mules, or at least our friends aren't.

So can people be cloned? Is there something unique about humanness that makes it impossible to clone?

The answer is both yes and no, as you'll see in this section. But before we go into details, let's review how an animal is cloned. (For greater details about the cloning process, turn to Chapter 6.)

To clone an animal, first an animal's egg is enucleated—the nucleus, containing its genetic material, is taken out of the egg. Then a cell is taken from the animal to be cloned, and the nucleus from that cell placed in the enucleated egg. That means that the egg now has the DNA of the animal to be cloned—in essence, the egg is a genetic clone of it.

BioSource

Most people would ban the cloning of human beings, but there is at least one organization devoted to making sure that human cloning is legal—and not just therapeutic cloning, but cloning human beings. The Human Cloning Foundation (www.humancloning.org) was set up to support human cloning. The Clone Rights United Front (www.clonerights.com) is also dedicated to supporting human cloning.

A jolt of electricity is sent through the egg with the new DNA in it. That starts the egg dividing. After it becomes an embryo, it is implanted into a female, and the embryo develops in the same way as a normal embryo would develop. Eventually, the animal gives birth. The resulting baby is a clone of the animal that donated the DNA.

This sounds like a simple process, but there are a great many steps along the way, and a great many dangers. As we explained in Chapter 6, cloning is still very much of a hit-or-miss proposition. Barely one percent of cloned embryos ever survive until birth. Considering that many cloning attempts never even make it until the embryo stage, the odds of producing a clone are even less than that.

Even when clones are born, however, they are often subject to diseases. Some are born very large, for example, and others are born with genetic defects. Clones are also subject to premature aging. For example, even though Dolly the sheep was young for a sheep she had arthritis, suffered from obesity, and was euthanized after a progressive lung infection.

Still even though there are problems with cloning, scientists are developing better techniques to ensure that a higher percentage of clones are born, and that the ones that are born are healthier. Cloning, after all, is still in its infancy.

BioFact

Scientists believe that clones suffer from premature aging because structures at the end of their chromosomes called telomeres are shorter than normal. Every time a cell divides, the telomere shortens. When the telomeres get very short, the cell has difficulty functioning properly and eventually can no longer divide, and so it ages and dies. So many cells are programmed to die after a certain amount of time. The DNA for cloning is generally taken from cells that have already divided many times, unlike an egg cell, which has yet to divide. So some scientists believe that one of the reasons for the health problems of clones may be that its cells are in essence prematurely aged starting from birth.

The Problem with Cloning Human Beings

One might expect that a scientist could clone human beings in the same way used for sheep. Take an egg from a woman, take out its nucleus, put in the DNA of the person to be cloned, turn it into an embryo, and implant it. Congratulations—it's a clone!

However, when it comes to human beings, things don't quite work that way. Humans are primates, an order which includes man, monkeys, apes, and animals called *prosimians*, which are lemurs and related animals. Scientists have tried many times to clone primates, and have been unsuccessful every time. Hundreds of attempts have been made, and at this writing, not a single one has succeeded—in fact, a single scientist tried 727 times to clone a monkey. In only 33 of the attempts did the cloned egg divide. Those 33 embryos appeared healthy, but when implanted, none have been able to develop properly. Other scientists have reported the same results.

The normal-looking embryos are in fact a genetic "gallery of horrors," said Tanja Dominko, a researcher who ultimately helped uncover why primates have not yet been able to be cloned. Almost all of the cells in the embryos did not have distinct nuclei that contained all of the animal's chromosomes. Rather than being found in the nucleus, the chromosomes were scattered unevenly throughout the cell, and so the cells couldn't function properly. In many cases, the cells didn't have the proper number of chromosomes. Strangely enough, some of the cells continued to divide, even though they didn't have proper nuclei.

So what's going on? Why should animals such as sheep be able to be cloned, but primates not be able to be cloned?

A team at the University of Pittsburgh School of Medicine discovered that when cells divide in a cloned primate embryo, they don't divide properly. The team studied what happened when cloned embryos of rhesus monkeys divided. Normally, when a cell divides into two, its chromosomes—which contain the genetic material—divide into two as well. Each of the two new cells gets a complete set of chromosomes from the original set of chromosomes, which divided.

BioFact

There is a chance that human cloning may be eventually banned worldwide. The U.S. Congress has been considering banning human cloning, as have the parliaments in several other countries, including the United Kingdom. The United Nations has been considering a worldwide ban on cloning as well. However, the ban has been held up because the United States and the Vatican want to ban all cloning, including therapeutic cloning, while many other countries, including France and Germany believe that therapeutic cloning should remain legal because of the significant health benefits that can be offered by it.

When chromosomes divide, they line up along a cell structure called a spindle, and are then pulled apart to the opposite ends of the cell. When the cell divides, the chromosomes then end up in two separate cells.

That's what *should* happen anyway. In the case of cloned primate embryos, however, something goes wrong. Rather than there being well-organized spindles which pull the chromosomes to opposite ends of the cell, there are instead "chaotic structures." This means that the chromosomes can't get pulled apart properly, and so they don't end up in the proper places. So all the chromosomes don't end up in the two new cells, and what chromosomes do show up may not end up in the nucleus because they haven't been pulled to the proper places.

The spindles don't form properly because two key proteins that normally help build and organize the spindle in primates—called NuMA and HSET—are missing from the cloned embryo. Without those proteins, the spindle can't be formed correctly.

Why are the proteins missing? As it turns out, the proteins are concentrated near the chromosomes of unfertilized eggs in primates. As we've explained, the first step in creating a clone is to remove the chromosomes of unfertilized eggs. So when you take out the chromosomes of the unfertilized primate egg, you also take out those vital proteins. And so the embryo can't divide properly.

In most other mammals, the proteins are scattered throughout the entire egg, and are not concentrated near the chromosomes. So taking out the chromosomes from the unfertilized egg of a sheep or a cow, for example, won't remove these vital proteins. That's why Dolly can be cloned, but so far a rhesus monkey can't be cloned.

What does this mean for human cloning? No one is quite sure. Some scientists say that it means that primates—including humans—will never be cloned. Others say that this is only a temporary hurdle, and that ultimately science will be able to leap across it. Only time will tell.

BioSource

The most comprehensive government look at cloning was done by the President's Council on Bioethics, which was formed as a result of President George W. Bush's decision to allow only limited funding for therapeutic stem cell cloning, and to allow no funding for human cloning. The report, all 350-plus pages of it, is available as a book and is published by Public Affairs under the title *Human Cloning and Human Dignity: The Report of the President's Council on Bioethics.*

Why Would Anyone Want to Clone Humans?

Even though human cloning is not possible now, it may become possible some day. And there are those who favor cloning human beings for a variety of reasons. We'll cover them in much more detail in Chapter 12. Meanwhile, here are some of the reasons that people believe humans should be cloned:

- **It can help infertile couples have children.** Those in favor of cloning say that infertile couples have the same rights as fertile couples to have children, and that with cloning, they will be able to have children who are genetically related to one of them.
- **To give birth to a child free of genetic disease.** If parents both have a recessive gene that carries a deadly or serious genetic disorder, there is a one in four chance a child of theirs will have that disorder. Cloning one of the parents would ensure that the child would be born free of the genetic disorder, because each parent has a healthy dominant gene that overrides the recessive gene.
- **To produce transplants for a sick child.** If a child were sick and needed a transplant—of bone marrow or a kidney, for example—a child could be cloned who would match the sick child's tissue, and that second child could then donate an organ to the first. The second child would still be able to live, and so both children would be healthy.
- **It would allow grieving parents to be consoled by a clone of a child who has died.** If a child died, and parents had some of the genetic material from the dead child, they could clone it, and so assuage their grief.
- **The right to reproduce by any means is a basic human right, and so cloning should be allowed.** Some people argue that no government has any

right to stop people from reproducing in any way they want, and so cloning is a basic human right.

BioFact

A couple in the United Kingdom conceived a baby with the same immune system genes as their older, sick child, in the hopes that stem cells from the baby's umbilical cord would help save the sick child. The child suffers from the genetic disease Diamond Blackfan anemia, in which too few red blood cells are produced. The parents traveled to the United States for treatment at the Reproductive Genetics Institute in Chicago. No cloning was involved, instead *in vitro* fertilization and gene testing were combined to pick out an embryo that closely matched the tissue of the child, and that embryo was then implanted into the mother and brought to term. As of this writing, it is not clear whether the stem cells helped the sick child.

Dangers and Problems with Human Cloning

Most people don't agree that humans should be cloned, and we'll cover those reasons in great detail in Chapter 13. In this section, we are not going to look at the moral or legal issues involved. Instead, we'll cover some of the medical dangers posed by cloning:

- **It isn't possible.** Currently, there is no way to clone humans, and it's not clear whether it will ever be possible. Because of the difficulties involved, the mere attempt to clone may prove to be dangerous.
- **Trying it will endanger the mother.** When animals are cloned, many eggs, not a single egg, need to be retrieved, because so many eggs never develop into healthy embryos. In order to get a woman to produce many eggs, powerful hormones must be used, which could endanger the woman's health. Additionally, no one knows whether a cloned embryo implanted in a woman may pose a greater danger than a normal embryo.
- **The first clones would be experiments—and humans shouldn't be subject to experimentation.** By the very nature of the procedure, cloning will necessarily be experimental, and human beings shouldn't be subject to experiments—especially a child yet to be born.

BioFact

A number of different surveys and polls have shown that Americans favor outlawing cloning human beings. A *USA Today*/CNN poll taken in 2003 found that 86 percent of Americans thought that human cloning should be made illegal.

- **Clones are frequently born sick or with genetic diseases and die young.** As we've explained throughout this book, clones are frequently unhealthy specimens. It would be unfair to bring a human being into the world, given the sizable likelihood he or she would suffer from a variety of diseases and conditions, and then die young.

> **BioDefinition**
>
> **Mitochondria** are the powerhouses of the cells, and provide the cell with the energy it needs in order to live and reproduce. Mitochondria can reproduce by themselves and have their own DNA.

- **There is no such thing as an exact clone.** As we explained in Chapter 6, there really is no such thing as an exact clone. Depending on the cell that donates the DNA to the clone, the genes in them may act differently from one another. Additionally, environmental factors affect which genes are turned on and off. And the genes of the cloned person will also be slightly different than the genes of the donor person, because the *mitochondrial* DNA will be different. (Mitochondrial DNA comes from the mother.)

The Least You Need to Know

- The UFO cult the Raelians claim that human beings are the result of genetic experiments carried out by aliens. They also claim to have cloned human beings through their company called *Clonaid.*
- Two doctors—Severino Antinori and Panos Zavos—claim to have cloned human beings or human embryos, but most scientists discount their claims.
- No primate has yet been proven to have been cloned, even though there have been hundreds of attempts.
- Primates have not yet been cloned because during the cloning process, vital proteins that the cells need in order to copy their chromosomes and divide properly are inadvertently taken out of the primate embryo. Those proteins aren't taken out of the embryos of other mammals.
- Those who favor human cloning say that it can help infertile parents give birth to children, and can help produce transplants for a sick child.
- Those who favor banning human cloning point out that there are many serious medical dangers that face the mothers of clones as well as the clones themselves.

Part 3 The Controversies Around Applying Cloning

Should animals be cloned so that their organs can be harvested for human transplantation? When, if ever, should a human be cloned? What does the Bible say about cloning? And the most burning question of all: Can computers and DNA grant us everlasting life?

In this part we'll cover the ethical, religious, and practical issues surrounding cloning. Part 3 will raise as many questions as it provides answers—and in fact, may well raise more of them. That's because cloning is so complex an issue touching so many parts of our lives, including what it means to be a human. So approach this section with an open mind, and possibly an open heart, and be ready to ask questions as well as find answers.

Chapter 10

Headless Humans and Genetic Pig Farms

In This Chapter

- Why organ donations are desperately needed
- How animal organs have been used in human beings
- Why pig organs are well-suited for human transplantation
- How transgenic pigs can be created and cloned for organ transplantation
- The pros and cons of cloning animals for organ transplants

One day you, or one of your friends or relatives, may be saved by a pig's heart—it may replace your diseased organ. That's the vision of a great number of researchers, anyway, who have been trying to clone transgenic pigs that can grow organs designed for human transplantation.

This sounds like the realm of science fiction, but it's very much real life. Already, pig heart valves are implanted into people. In addition, researchers have cloned batches of pigs that may prove to have organs suitable for human transplantation.

In this chapter, we'll look at how and why such transplantation might be done, and some of the issues surrounding the technique.

Who Needs Spare Organs?

To a certain extent, medical science has progressed so far that it's able to treat the body like a machine. When a part goes bad and can't be repaired, just throw it out and replace it with a new one.

BioSource

For information on organ donation or on becoming an organ donor, go to the Coalition on Donation website at www.shareyourlife.org; to the United Network for Organ Sharing at www.unos.org; and the federal Department of Health and Human Services Organ Donation website at www.organdonor.gov.

That's the idea behind organ transplantation. Diseases can so damage organs that they simply can't function properly any longer, and so they are removed, and replaced with a new one. The entire organ can be removed and replaced or tissues from it can be removed and replaced.

Although organ transplants frequently involve major operations, in many cases they are relatively routine and the survival rates are quite high.

But there's one problem—there simply aren't enough organs available for everyone who needs one. In fact, an increasing number of young men in poor countries agree to "donate" their kidneys in return for large cash payments from rich patients and their doctors. Consider these facts from the Coalition on Donation, a nonprofit group that promotes organ donation:

- Over 80,000 people are currently waiting for life-saving organ transplants.
- Every day 17 people die because organs aren't available for transplantation.
- Every 13 minutes, another person's name is added to the list of people waiting for an organ transplant.

Organ donation is voluntary, of course, and even though many private groups, the federal government, and state governments try to publicize and promote organ donation, the problem is not getting any better. In fact, in future years it may well get worse. As the population ages, a higher percentage of people are likely to need organs. And as medical science continues to make advances, it will learn new ways to keep people alive who may previously have died, and who could benefit from an organ donation.

BioFact

Someone with a transplanted organ can often live as healthy and active a life as if they had not had a medical problem. Snowboarder Chris Klug, for example, suffered from PSC (Primary Sclerosing Cholangitis), a degenerative liver condition. Without a new liver, he would have died. A donor was found and four months after the transplant operation he was participating in the World Cup snowboarding circuit. Eighteen months later, he won a Bronze Medal in the Olympics.

So what to do? In the eyes of many researchers and doctors, the answer is simple: Use animal spare parts instead of human spare parts. That's what we'll look at in the next section.

Of Pigs and Men

The idea of using animal organs in human beings may sound far-fetched and very twenty-first century, but in fact for millennia mankind has imagined human-animal hybrids. The Greeks and Romans, in particular, were big on the idea—their mythology is strewn with them. Centaurs were human-horse hybrids, while satyrs combined a goat and a man. A sphinx was a combination of woman and lion—with wings thrown in for effect. And many cultures have a fascination with human-fish hybrids, like mermaids and mermen.

Of course the Greeks and Romans didn't go into details about how those human-animal hybrids came into being—most likely Zeus had one of his many gargantuan temper tantrums and used a bit of Olympic magic to do the trick. We humans aren't so lucky, however, and so we have to use modern medical technology.

The idea of using animal organs for transplants is not that new—experiments using them have been going on for four decades, starting in 1963 with the transplantation of chimpanzee kidneys into human beings. The technique of transplanting animal organs into humans is called *xenotransplantation*. (For more details about the history of animal organ transplants, see the section "A Short History of Xenotransplantation" later in this chapter.)

BioDefinition

The preface "**xeno**" means stranger or foreigner in Greek.

But there's a major problem with xenotransplanation. The human body recognizes the organ as foreign to it, and so the immune system attacks it ferociously. In fact, within a few minutes of a transplant, the immune system targets the organ and begins to destroy it. This is an example of a good process gone bad—normally the foreigner would be an invading germ but the immune system doesn't know it's a beneficial organ instead.

To get around the problem, doctors use drugs that suppress the immune response. That way, the body won't attack the foreign organ. However, this isn't always effective and the body frequently still attacks the organ. Suppressing the immune system has a very severe unintended side effect: The body can't fight off infections. Frequently, in tests when animal organs were implanted into humans, the patients were killed by infections that the body couldn't fight off, and not by failure of the organ. These infections are normal ones that the body would usually kill; the infections typically do not come from the animal organ.

Doctors have implanted a variety of animal organs into humans, including those from chimpanzees, baboons, and pigs. But these days, when it comes to xenotransplantation, pigs are where the action is.

There are a number of reasons for using pigs. One is what you might call "form factor"—pig organs tend to fit neatly in the human body as they are of a comparable size to the organ that they're replacing. Pigs are also inexpensive and easy to breed and raise, unlike primates such as baboons and chimpanzees. Pigs are better than primates for another reason as well: Primates are similar enough to human beings so that the risk of *zoonosis*—the dangers of viruses "jumping the species barrier" from primate to human—is rather high. In other words, a primate virus is more likely to thrive in a human being than a pig virus, and therefore could be more dangerous. For example, it is generally believed that AIDS originated in a chimpanzee and then made the jump to human beings who ate diseased chimps.

BioDefinition

Zoonosis is when an infectious agent makes the jump from one species to another.

How Cloning and Genetic Engineering Help

Genetic engineering and cloning can help with the major problem of using animal organs for human transplantation—using cloning and gene-implanting techniques, the human body can be fooled into thinking a pig organ is a human organ.

There are many reasons why pig organs are rejected but a primary hurdle is the presence of a sugar called gal on the surface of the pig's cells. The human immune system recognizes the gal as a foreign substance, perhaps associated with bacteria, and begins attacking the pig cells. Gal is created by a pig enzyme called alpha-gal (for those who prefer formal names, it's enzyme alpha-1,3-galactosyltransferase). A specific pig gene,

GGTA1, is responsible for creating this enzyme. So if the gene could somehow be turned off, researchers hoped, the enzyme wouldn't be created, and it in turn wouldn't create gal. And then, the hope is, the human body would be less likely to reject the pig organ.

BioFact

The first transgenic pig, named Astrid, was created in 1992 by inserting a small amount of human DNA into a fertilized pig egg.

Scientists have managed to create pigs that lack the GGTA1 gene, using a combination of genetic engineering and cloning techniques. They first took DNA from a pig and eliminated the gene from its site on one of the chromosomes—knocked it out in scientific parlance—then implanted the DNA into a pig's egg whose DNA had been taken out. They then forced the egg to divide, and implanted the resulting embryo into a sow (a female pig). The resulting cloned "knockout" pig lacked the gene on that chromosome.

However, as you've learned throughout this book, animals have two copies of every chromosome, not one. The gene had only been knocked out in one of those chromosome pairs and so the work was only partially done. Scientists then took DNA from the pigs they had cloned, knocked out the remaining copy of the gene from the other chromosome, then cloned a new pig using that DNA. In 2002, the "double knockout" piglet was born, a pig that didn't have the GGTA1 gene, and therefore didn't have gal on its organs.

A great deal of testing has to be done before these pig organs can be transplanted into humans. Preliminary tests using the pigs with the double knocked-out gene, however, have been encouraging. Cells taken from the knock-out pigs were implanted in mice, and the mice didn't exhibit the extreme immune response that they do when cells from normal pigs are implanted in them. The next step will likely involve testing chimps or other primates, and then human trials will begin.

BioFact

Pig heart valves are commonly used as a replacement for human heart valves that have become damaged. No genetic engineering is used for the valves. They are treated with a chemical called glutaraldehyde that prevents the human body from rejecting the valves. Unfortunately, it is not possible to transplant whole treated pig organs because they would no longer work.

Beware of Zoonosis

There is a fly in the ointment in all this—or to be more precise, a virus in the pig tissue. Pigs are *less* likely than primates to have infections that can be passed along to humans, but that doesn't mean the possibility doesn't exist. In 1997, scientists discovered that a virus that infects pigs, called porcine endogonous retroviruses (PERV), could also infect human cells in culture. That doesn't necessarily dictate that PERV can infect living humans, but no one is sure whether they can or not, and no one wants to take a chance.

PERV apparently doesn't cause any symptoms in pigs, and they don't know what would happen if the virus were to infect humans. PERV is particularly problematic, because it incorporates itself into the pig's DNA, and is passed down through succeeding generations. That means that even if you raised a pig in a thoroughly sterile laboratory, it could still carry the virus because it could have gotten it from its parents' chromosomes. However, there are standard tests to find out if this has happened.

Therefore, before cloned pigs can be used to grow organs for humans, a method has to be found to ensure that PERV, or any other infectious agents, can't make the jump from pig to human.

CAUTION

BioWarning

When viruses cross species they can become very dangerous. This is why scientists are so careful these days when it comes to animal transplants—they don't want to create the next pandemic. Many emerging diseases are apparently caused when viruses make the jump from species to species. AIDS is the most deadly one, but Severe Acute Respiratory Syndrome (SARS) is believed to have jumped from another species to man in the southern Chinese province of Guangdong.

One of the groups that was able to clone pigs with double knocked-out genes, Immerge BioTherapeutics, Inc., claims that the pigs they cloned "seem incapable of transmitting PERV to human cells in culture," which is good news. But not good enough, because there's no definitive proof yet that it cannot happen. So Immerge and other companies have been studying PERV, and Immerge says that it has discovered the method by which PERV infects cells. The company is hoping to use that knowledge to figure out ways to ensure that PERV can't be passed along to humans.

A Short History of Xenotransplantation

The idea of using animal organs for humans is nothing new, and predates genetic engineering by decades. In fact, the first time an animal organ was implanted in a human being was back in 1963. Here's a brief look at some of the notable experiments that involved transplanting animal organs into humans up until the dawn of the twenty-first century:

- **1963-1964** Thirteen chimpanzee kidneys are transplanted into human beings at Tulane University in New Orleans. Twelve of the patients die within 60 days, but one patient survives for 9 months.
- **1964** Six baboon kidneys are transplanted into humans at the University of Colorado. Patients died within 19 to 98 days.
- **1977** Dr. Christian Barnard, who had performed the first human heart transplant, transplants a chimpanzee and a baboon heart into humans, in the hopes that they will keep the patients alive until human hearts can be found for them. The patient who received the baboon heart dies after six hours, and the one who received the chimpanzee heart survives for four days until the heart is rejected.
- **1984** A baboon heart is transplanted into a newborn infant, called "Baby Fae." She survives for 20 days before the heart was rejected.
- **1992** A baboon heart is transplanted into an AIDS patient who also suffers from hepatitis B. Doctors believe that baboons are resistant to hepatitis B, which is why they chose to transplant a baboon heart. The patient survived for 70 days.
- **1992** A pig heart is transplanted into a human by a Polish surgeon. The patient died within 24 hours.
- **1995** Clinical trials are conducted into treating patients with Parkinson's Disease by using nerve cells from fetal pigs.
- **1995** The Federal Food and Drug Administration (FDA) approves a proposal to test using transgenic pig livers in human beings as a "bridge" until human donors can be found for the patients.

> CAUTION
>
> **BioWarning**
>
> Seven years after the baboon heart was transplanted into the AIDS patient, researchers examined stored blood and tissue samples from the patient and discovered a baboon virus in them. The researchers could not determine whether the virus had actually infected the human being, or whether the human tissue samples were just contaminated with baboon cells.

- **1997** The discovery is made that viruses that infect pigs, called *porcine endogonous retroviruses* (PERV), can infect cultured human cells. Because of this, the FDA stops all clinical xenotransplanation trials until a technique is developed for detecting low levels of PERV infection. The technique is developed and the ban is lifted in 1998.
- **1999** The FDA bans using nonhuman primate organs for xenotransplantation into humans warning that the risk of cross-species infection is too high.

- **2000** Transgenic pigs with "knockout genes" are cloned for the first time.
- **2000** A study in the journal *Science* finds the first evidence of cross-species transmission of a retrovirus, when mice develop PERV infections after being transplanted with pig pancreatice cells.
- **2001** In a clinical trial involving Parkinson's Disease, patients with the disease received transplants of fetal pig cells. Those patients showed no improvement compared to a control group who did not receive the transplants.

Should Animals Be Used as Organ Farms?

The potential benefits of using animal organs for transplantation into humans is obvious: It can help save lives—potentially thousands of lives. However, there are many potential drawbacks as well, ethical, religious, and medical. In this section, we'll look at the potential problems and ethical and religious concerns with using animal organs.

BioFact

There is a long history of using pigs in treating humans. Since the 1930s, pig insulin has been used to help diabetics, because human insulin was so difficult and expensive to produce. Only in 1982, with the introduction of genetically engineered insulin, did it stop being in widespread use. Pig skin grafts have been used for burn victims as well.

Medical Concerns over Xenotransplantation

The major potential medical problem with transplanting genetically engineered pig organs is whether a virus or some other disease will cross the species barrier and infect humans. As explained earlier, the primary concern is PERV. But pigs can be infected with a variety of other viruses, bacteria, and parasites that can be passed to humans and infect them. These include rabies, influenza A, Eastern equine encephalitis, listeria, and others. With proper controls, one can be virtually certain that the organs to be transplanted won't be infected with these agents.

A larger concern is over viruses and bacteria that we don't know about. Could there be some kind of killer bug that we don't know about that could cross the species barrier?

Could There Be Better Use of Societal Resources?

Some people contend that the amount of money spent on research on xenotransplantation is wasted—that the problem could more easily be solved by spending the money to get people to donate their organs. They point out that most people do not donate organs, and that if most people did, there would be an organ surplus rather than the situation we now have in which there is an organ shortage.

Others believe that the ultimate solution will be in the use of stem cells to grow new organs. (For more information about stem cells, turn to Chapter 8.)

But those who favor xenotransplantation note that despite all the money and resources spent by government agencies and private groups trying to get people to donate organs, the rate of donation is still very low. And they note that using stem cells to grow organs is even more futuristic than is the use of pig organs for humans. For example, the Massachusetts biotech company Advanced Cell Technology does research in both breeding transgenic animals for organ transplantation and in stem cell research. According to a *Frontline* report, Robert Lanza, vice president of Medical and Scientific Development for the company, believes that animal-to-human transplants is not that far away, but that the use of stem cells to grow organs is a better long-term solution.

Religious Issues

There are some in religious groups who believe that both genetic engineering of animals and using animal organs for humans is wrong because it lets man, in essence, act as God. And some say that it upsets the balance of nature, and so oppose it for that reason. For example, the United Church of Canada says in a statement, "Xenotransplantation clearly transgresses the boundaries of species integrity that have evolved over the course of Earth's history, creating 'in-between' species that are neither fully animal nor fully human. We must admit with humility that this could have dangerous, even catastrophic effects, as the risk of disease crossing the species barrier illustrates."

BioFact

At least one church has come out with statements condemning the use of xenotransplantation. The United Church of Canada has called for a ban on all xenotransplants and has asked that governmental and corporate funding for the procedures be ended.

Animal Rights Issues

Worldwide there is a large and very vocal animal rights movement and, as you might expect, they are not pleased with the idea of xenotransplantation (or with genetic engineering in general, for that matter).

Their concern is a basic one. They believe that animals have rights and should not be subject to being raised merely so that they can be slaughtered and their organs harvested. They claim that pigs, for example, are highly intelligent and sensitive and that raising them for organs is cruel. They note that pigs are already being raised for meat, but they disagree with that as well. In the words of the Campaign for Responsible Transplantation, "Policy-makers in the United States and elsewhere have decided that it is 'ethical' to use pigs in xenotransplants because pigs are killed for food. But two wrongs do not make a right."

BioSource

One organization that has been formed specifically to fight against xenotransplantation is the Campaign for Responsible Transplantation. It is lobbying for a total ban on xenotransplantation. Find it on the web at www.crt-online.com.

The groups also argue that the mere creation of transgenic animals is cruel, because some of them are born with physical abnormalities including arthritis, muscular weakness, and defective vision. They argue that genetic engineering in animals and xenotransplantation by its very nature requires a great number of animals to be experimented on and killed.

How About Headless Humans?

Ah, the general press. What would we do without it? You may every once in a while see a quote or an article pop up predicting vast farms of headless humans. A lot of these predictions come from a single source, Dr. Patrick Dixon, the author of *The Genetic Revolution* and head of the organization Global Change. Dixon predicts that colonies of cloned headless humans could be used essentially as spare parts factories.

No serious scientist expects headless human organ farms, and no government would allow it. So rest easy tonight—the main thing you need to fear is tabloid headlines.

BioFact

Humans can't survive without a head, but in the 1940s there was a famous headless chicken, called, not surprisingly, "Mike the Headless Chicken" or "Miracle Mike." Mike was destined to grace someone's dinner table, but the hatchet aimed at his head went a bit awry, and left his neck and brain stem intact—intact enough so that Mike lived for several years after, being fed through a hole in his esophagus.

The Least You Need to Know

- Over 80,000 people are currently waiting for life-saving organ transplants, and every day 17 people die because organs aren't available for transplantation.
- Using primate organs for transplanting into humans can be dangerous because there is a high risk that viruses and other agents may infect humans.
- Pig organs are the right size for implanting into humans. Pigs are also easier and less expensive to raise than primates.
- Humans reject pig organs because of a sugar called gal that is present on the surface of the organs, but bioengineering may be able to create transplantable organs that lack the sugar.
- Trials using cloned transgenic pigs for human organ transplants are probably years away.
- Some animal rights groups and religious groups would ban the use of animal organs for transplantation.

Chapter 11

Whose Life Is It, Anyway?

In This Chapter

- Why are there patents—and what do they protect?
- How patent law applies to bioengineering, cloning, and genes
- How bioengineered, oil-hungry bacteria changed patent law forever
- The case of the bioengineered, patented mouse that roared
- The pros and cons of patenting life and genes

Everybody knows something about patents. When you invent a new machine, or a new procedure, or some kind of new device, you can get a patent on it and, if things go well, make some money when it's mass produced.

But you may not know that you can get a patent on a life-form. There's an old saying, build a better mousetrap and the world will beat a path to your door. You can update it to read: build a better *mouse* and the world will beat a path to your door. Because as you'll see in this chapter, a new bioengineered mouse is only one of the many thousands of living things that have received a patent in the last 20 years.

In fact, even individual genes or parts of genes have been patented. This is one of the most controversial aspects of bioengineering and cloning. So to find out about how to patent life, and the issues around it, dig into this chapter.

A Brief Primer on Patents

As you'll see later in this chapter, the range of life and bio-stuff that you can patent is quite astounding—everything from gene segments to genes to artificially created life and transgenic plants and animals, and even the entire genetic heritage of an entire nation.

To make sense of why all that is possible, though, you'll need a brief primer on patents. Don't go to sleep on us—this is actually pretty interesting stuff.

BioFact

The first modern laws that gave inventors exclusive rights to their inventions for a specified period of time are believed to have been put into effect in Italy in the fifteenth century. In the United States, before the Constitution was ratified, most states had their own patent laws. The U.S. Constitution gave to Congress the power to regulate patents in Article 1, Section 8, Clause 8, which reads in part: "Congress shall have power … to promote the Progress of Science and useful Arts by securing for limited Times to Authors and Inventors the exclusive Right to their respective Writings and Discoveries."

A patent is an agreement that an inventor makes with the government in which the inventor has the sole rights to use his invention, in return for completely disclosing information about the invention. The inventor can in turn sell those rights to other people or companies. The patent is good for 20 years, and then anyone can use the invention. So that means that inventors want to get as much money out of the patent as soon as they can, because they lose their monopoly after 20 years. Often they have far less than 20 years because the patent is sometimes just a roadmap and actually producing the product can take years.

Patents serve two primary purposes: To help advance technology, and to protect the inventor so that he can enjoy the fruits of his invention.

To receive a patent, an inventor must show that his invention meets three primary criteria:

- **It has to be truly new and "nonobvious."** So it can't have been previously invented, and it can't be obviously based on the work of others.

- **It has to be useful.** So the inventor has to show that it has a real-world use of some kind.
- **The patent application has to describe the invention in enough detail so that someone can reproduce it.** If that detail is missing, the patent won't be approved. Still, only the inventor has the rights to use it for 20 years, although he can sell the rights.

BioFact

The first U.S. patent was issued to Samuel Hopkins of Philadelphia in 1790, who invented a new way of manufacturing potash, which was used in making gunpowder, soap, and glass. Unfortunately, he was a better inventor than he was a businessman. He was never able to cash in big on his invention and died in obscurity. A Quaker, he was even ostracized by fellow Quakers who believed that excessive risk-taking disturbed one's "tranquility of mind." Making matters worse, for years afterward historians credited a different Samuel Hopkins for being granted the first U.S. patent.

There are a couple of other conditions as well. A description of the invention can't have been published before the patent is applied for. In addition, the invention cannot have been on sale or in general use in the United States for a year before the patent application was made.

How Does This Apply to Cloning and Bioengineering?

The original U.S. patent laws were written by Congress in 1790, about two centuries before the biotech revolution. So our founding Congressmen certainly didn't have cloning, DNA, or transgenic plants and animals in mind when they wrote the law.

So how do the patent laws apply to biotech? As you'll see in the rest of the chapter, the issues are pretty tricky, but there are some general rules that apply. First, there are three kinds of biotech inventions that are patented:

- An invention based on a useful, just-discovered, isolated gene or protein. Therefore, a naturally occurring gene or protein can only be patented if it is isolated and its properties suggest a useful commercial purpose.
- A new way of treating a disease using a specific gene or protein. For example, if someone has a patent on a gene or protein, another inventor can get a patent on a new way to use that gene or protein, as long as the second inventor has found a new way that the gene or protein can be used.

- A laboratory procedure that is useful for isolating genes or proteins. The patents explaining how to "splice genes" so that proteins like human insulin can be made have generated huge sums for the inventors and their universities.

As with any other inventors, biotech inventors have to meet the three basic criteria outlined earlier in this chapter in order to receive a patent. For the novel and non-obvious criteria, for example, the inventor has to know the precise chemical structure of the gene or protein he is trying to patent—and if the structure has already been precisely described by someone else, he can't get the patent.

BioSource

If you're interested in finding out information about patents, or want to search through the entire database of U.S. patents, go to the U.S. Patent Office's website at www.uspto.gov.

For the useful criterion, the inventor has to specify its real-world practical use. Simply describing the precise chemical structure of the gene or protein isn't enough. He also has to detail how it will be used, for example, as a treatment for a specific disease.

The Tale of the Hungry Bacteria

Everyone knows that oil and water don't mix, but how about oil and bacteria? If you're a certain kind of scientist, you see a connection between oil and bacteria that the rest of us don't see. You want recognition for the research, and if there is money to be made, you want to make sure that you and your colleagues reap the benefits of your work.

In 1971, a General Electric scientist named Ananda Chakrabarty genetically engineered a bacteria that would literally eat oil. He had noted that certain strains of bacteria had a taste for oily matter and figured that he could put that habit to good use. So he took genetic material from three different kinds of bacteria and put them into a fourth bacteria. He came up with a kind of superbug, a type of Pseudomonas bacteria that ate oil for lunch (and for breakfast, dinner, and snacks in between). The oil-eating bacteria had never existed before in nature. It was a thoroughly human invention. The idea behind it was a very practical one because in theory, the bacteria could help clean up disastrous oil spills.

So Chakrabarty applied to the U.S. Patent Office for a patent for the new life form. They turned him down flat, saying that bacteria were products of nature and because they were living things, they could not be patented. Chakrabarty disagreed, and appealed the decision. The case ultimately made its way to the Supreme Court nine years after he had created the bacteria.

In a close 5–4 decision, in a case titled *Diamond* v. *Chakrabarty* the court made the most momentous decision of the biotech era. It ruled in 1980 that life forms were, in fact, patentable, and ordered the Patent Office to award Chakrabarty a patent (U.S. Pat. No. 4,259,444 to be exact). Its wording on the ruling could not have been broader or clearer. It ruled that a patent can be granted on "anything under the sun which can be made by man." It went on further to say that whether something could be patented had nothing to do with whether that thing is living or not. Instead, it revolved around "whether living products could be seen as 'human-made' inventions."

BioFact

While Chakrabarty's patent was the first granted for a new life form, the U.S. Patent Office has long given patents on natural products, in cases where procedures are described to isolate them in a pure state. For example, in 1873 Louis Pasteur was granted a patent on yeast which was "free from organic germs of disease, as an article of manufacture." He had been the first person to create disease-free yeast. And in 1911, a patent was granted for the creation of purified adrenaline. Even though adrenaline exists naturally in the human body, it had never been isolated outside the body and purified.

In Chakrabarty's case, the bacterium was clearly a human-made invention, because it does not exist in nature—it was created by genetic engineering.

That ruling opened the floodgates to countless cases of patenting new life-forms, genes, and other products of bioengineering.

BioWarning

Chakrabarty ultimately received his patent, but his oil-eating bacteria have not yet been tested in nature. He was able to prove in the lab that it ate oil, but it has not been released into the environment. If released, it could potentially combine with pathogenic bacteria that are its relatives, with potentially dangerous results. So it remains a laboratory bug.

The Mouse That Roared

The patent issued for oil-eating bacteria was a milestone because it recognized that bioengineered life-forms could be patented. But it was more important for its legal than its scientific implications, because other scientists were not rushing to use the oil-chomping Pseudomonas bacteria in their research.

From a scientific and research point of view, a much more far-reaching patent—and an extremely controversial one—came via the hallowed halls of Harvard University. Dr. Philip Leder at the university developed what is called the Oncomouse—a mouse that was genetically modified by giving it an oncogene, a gene that would cause cancer. This gene is passed down from generation to generation of mice, so that each succeeding generation will develop cancer. And not only will the mice develop cancer, but they will develop cancer in a predictable way.

Now, this may bring to mind an old *Saturday Night Live* skit in which mice were dressed in polyester leisure suits as a way to test the cancer-causing properties of polyester. But the development of the mouse was a significant breakthrough in cancer research. The genetic makeup of mice is similar in many ways to that of humans. Additionally, mice are easy and inexpensive to breed and keep so they are widely used in research, and frequently in cancer research. Having a mouse that will consistently develop cancer in a predictable way means that cancer researchers can much more effectively test anti-cancer drugs and agents, and study how cancers develop.

Dr. Leder had developed the mouse with funding from the DuPont Chemical Company. So the patent was issued to Harvard University, which in turn gave exclusive licensing rights to DuPont.

BioFact

The Oncomouse develops cancer because it was bioengineered to have a mutation in the MYC gene, which encodes proteins that regulate cell growth and cell differentiation. Because the mutated MYC proteins can't function properly, the mouse develops cancer when some cells grow wildly out of control.

That's where the trouble comes in, and it illustrates one of the reasons that many researchers are concerned about patenting life forms. DuPont, some researchers complained, was charging exorbitantly high fees for the rights to use the mice. Additionally, said others, the paperwork involved in gaining rights to use the mice was onerous.

In response, DuPont in signed an agreement with the Public Health Service of the U.S. Department of Health and Human Services in 2000 that was supposed to have given academic researchers who use federal funds for research the rights to use the mice without paying a fee to DuPont.

However, since that agreement was signed, researchers have complained that DuPont has gotten even more aggressive about trying to collect fees for use of the mouse. In fact, one of the scientists instrumental in putting together that agreement with DuPont has complained bitterly about it.

It's now at the point where some researchers say that DuPont is actively impeding cancer research. They claim that DuPont imposes unrealistically strict conditions on university researchers, and that DuPont charges too high a fee to companies that want to use the mouse for cancer research. Most alarmingly, they say that DuPont has interpreted the patent exceedingly broadly. They claim DuPont is trying to charge fees to anyone doing research with cancer-prone mice, even if those mice aren't the Oncomouse covered by the patent. If another method were developed to create a cancer-prone mouse, DuPont is trying to charge a fee to them, say the researchers.

Because of this, researchers at a variety of major universities have either refused to sign DuPont licensing agreements, or delayed signing them. These universities include the Massachusetts Institute of Technology and the University of California system. Additionally, the Memorial Sloan-Kettering Cancer Institute in New York refused to sign. Harold Varmus, President of the Memorial Sloan-Kettering Cancer Institute, was former director of the National Institutes of Health, and helped put together the DuPont agreement. He told the *Boston Globe* that he was displeased with the way that DuPont was carrying out its end of the agreement. The *Boston Globe* also noted that Varmus and other scientists were told by drug companies that DuPont was charging hundreds of thousands of dollars for the right to use cancer-prone mice in testing.

BioFact

Not all of the countries of the world agree that a patent should be granted to Harvard for the Oncomouse. The Supreme Court of Canada, for example, has ruled that the Oncomouse cannot receive a Canadian patent. And even though Harvard received a patent for it in Japan and some European countries, those rights are still under dispute.

The Right to Own Genes

While you may or may not agree that genetically modified creatures like the Oncomouse and the Pseudomonas bacteria should be allowed to be patented, you probably can follow the logic of those who claim they should be. If someone hadn't created them, they would not exist, and so they are in that sense an invention.

Harder to understand, though, is the right to patent a gene. And we're not talking about genes that have somehow been artificially created or modified, and so otherwise wouldn't exist in nature. We're talking about genes that naturally occur in animals, including humans. And it's not only genes that can be patented, but segments of genes as well.

BioFact

At the end of 2001, the patent office had issued more than 6,500 patents for genes and DNA sequences that code for proteins. More than 1,300 of those patents were for human genes and DNA sequences. There have been more than 20,000 applications for genes and DNA sequences.

Why can this be done? Remember our patent primer? If a gene or protein can be isolated outside the body and shown to be useful as a drug, in testing or in some other way, it can be patented. That's as long as it's "non-obvious" how to go about doing that.

In the rest of this section, we'll review some of the more famous—and infamous—cases in which genes have been patented, and how those patents are being used. Later on in this chapter, we'll go into details about the controversy surrounding the patenting of genes.

The Sad Story of the Hairy Spleen

One of the oddest and more controversial cases of patenting genes and cells has to do with cells taken from a cancerous 22-pound spleen—and the spleen's owner had no idea that his cells were going to be patented.

John Moore, a Seattle resident, suffered from a rare form of leukemia called *hairy-cell leukemia*. He traveled to the University of California in 1976, where he underwent a variety of tests. According to the court records of the case, after taking these tests, Moore's doctors "were aware that certain blood products and blood components were of great value in a number of commercial and scientific efforts and that access to a patient whose blood contained these substances would provide competitive, commercial, and scientific advantages." They did not, however, tell Moore that.

Moore's cancerous spleen was removed in October 1976, and he traveled between his home and UCLA a number of times between then and 1983 at the request of his doctors. Each time he visited, blood tests and other tests were performed, and Moore was told that the tests should only be done by his doctor and only at UCLA. What his doctor didn't tell Moore was that they were "conducting research on Moore's cells and planned to benefit financially by exploiting the cells and their exclusive access to the cells by virtue of … the ongoing physician-patient relationship," according to the court.

The doctors isolated a cell line from the spleen that had the potential for helping in cancer therapy. The doctors applied for a patent to the cells without telling Moore, and they received the patent. After receiving the patent, one doctor received 75,000 shares of stock in the firm Genetics Institute, which was getting the rights to the patent. The doctor also got a consulting contract, and both the doctor and the

University of California system received payments of $440,000 over the next several years. There may have been other payments as well, but they weren't covered in the court filings.

BioFact

A British woman tried to patent herself as a way of protecting herself from "genetic exploitation" in her words. She applied for a patent from the United Kingdom's Patent Office. She told the Reuters news agency, "It has taken 31 years of hard labor to invent myself. It may sound odd, but I want to make sure I can protect myself from unauthorized exploitation, genetic or otherwise .... There is a kind of greedy, unpleasant atmosphere around the mapping of the human at the moment. I want to have sole control over my own genetic material."

In fact, tumors removed in life-saving operations are often used for research purposes, sometimes years later, but who should "own" them—the patient or the hospital? Moore sued, claiming that he should have the rights to his own cells. He lost. The California Supreme Court ruled that he no longer had any rights to his cells after the cells had been removed from his body. Moore sued on a related charge and eventually settled out of court for an undisclosed sum of money.

Owning All the Genes in Iceland

It may seem odd that someone could patent someone else's genes and cell lines. But stranger still is the agreement that the government of Iceland made with the firm deCODE Genetics in which the government gave the biotech company a 12-year monopoly on the commercial use of, essentially, all the human genes in Iceland.

The company is compiling a database containing the genetic information of everyone in Iceland, although the information won't be able to be traced back to individuals. It is tracking the links between specific genes and diseases. And those who want to "opt out" of the database can do so. The firm will then have exclusive rights to the use of that database, and to patent any genes or similar material based on the database, for a period of 12 years. The government received an undisclosed payment in return and the people of Iceland will also receive any drugs created from the results of the research free of charge.

Why Iceland? Biotech firms are looking for a stable genetic population, and Iceland is isolated, inbred, and fits the bill. Additionally, Icelanders are a detail-oriented bunch, and they have been saving their health records since 1915, a big aid to researchers. Icelanders are also obsessive about their family trees, so there is an abundance of this kind of information available—again, a boon to researchers who might want to trace genetics through generations. Iceland has an ideal population for finding genes associated with diseases that have been passed down through families.

The deal has been controversial in Iceland as well as in the rest of the world, but remains in effect.

BioFact

The deCODE project has already born some fruit. For example, the company claims that it has found a gene that is linked to high blood pressure and that might some day lead to medicines to better control the condition. It also claims that it discovered a gene with links to asthma and two others linked to allergies. They plan to use that knowledge to develop a new diagnostic test to help treatment of the conditions. Even more recently, the company identified a gene that it found was associated with increased risk for stroke.

Why Genetic Patents Are Good

Those who favor patenting life forms, genes, and genetic material say that it aids in the search for better medical treatments. They point out that medical research takes a significant amount of money, and that frequently research leads to dead ends. The only way that research can be funded, they maintain, is if companies know that there will be a financial payoff for them in the long run. A patent insures that financial payoff because the companies will benefit from its exclusive rights to that new technology.

BioFact

The University of Hawaii made patent history by filing for a patent to a gene that causes a genetic disorder—and voluntarily sharing the patent with a patient support group for the disease. The patent was for a gene that causes the disease pseudoxanthoma elasticum, or PXE. The disease can cause hardening of the arteries, blindness, premature aging, and gastrointestinal bleeding.

They also say that, just as the Supreme Court ruled, there is no difference between a nonliving and a living invention. Therefore, inventors who invent living things should have the same rights as those who invent nonliving things.

Finally, they note that a patent requires that detailed information about the invention be published, and that helps spur more research as the information is publicly available to anyone, free of charge.

Have Biotech Gene Patents Gone Too Far?

Not everyone agrees that these kinds of patents are a good thing. There are those who believe that biotech patents have gone too far. Many are uncomfortable with the idea of patenting genes, because they are simply naturally occurring and no real invention is involved. This is different from the patent for the oil-eating bacteria we just discussed because that was a brand new invention. Biotech companies argue that if they couldn't patent a gene that is an obvious drug target then there would be little incentive to go forward with expensive research.

Such complaints don't come only from the "usual suspects" such as environmental groups and various nonprofit organizations. Many complaints come from within the scientific community and from biotech firms as well. These groups worry that patent fever is slowing down important research that could otherwise lead to the cure and control of many diseases. In this section, we'll take a look at the main concerns.

Less Free Exchange of Ideas

Good science requires the free exchange of ideas—and there are those who contend that biotech patents cut down on that exchange, and harm all of us in the long run.

Harold Varmus worries about that. He testified before Congress about the gene patent issue in July of 2000, and noted that patenting rewards the obvious work of sequencing DNA, which can be done much more easily than in the past, and downplays the exact function and utility of the genes. As a result, he contends, universities fund costly offices that protect their intellectual property and stop scientists from freely sharing their work with others. The open exchange of ideas so necessary to science has been inhibited, he said.

Overbroad Patents

Some scientists contend that biotech patents for genes can be overbroad. For example, at the same hearing at which Varmus testified, Jon Merz, professor of Bioethics at the University of Pennsylvania, testified that patents are interpreted so broadly that they cover a great many tests and clinical work that shouldn't be covered. As a result, he maintained, physicians and researchers do not carry out important genetic tests because of the fear of patent infringement.

> CAUTION
>
> **BioWarning**
>
> Professor Merz backed up his testimony with research. He and several colleagues surveyed 74 laboratory physicians and found that because of patents 25 percent had abandoned a clinical test they had developed, and 48 percent had not developed a clinical test.

Patents Stifle Innovation

Because patents can cut down on the free exchange of ideas, innovation is stifled by them, say critics. This is ironic, they note, because patent law is designed to foster innovation, not squelch it.

Regis Kelley, executive vice chancellor of the University of California at San Francisco, warned at a hearing before the Federal Trade Commission, "The proliferation of gene patents does create something of a minefield that could inhibit the commercial development of new medicines."

Even some in the biotech industry believe that gene patents are harming medical progress. Barbara Caulfield, general counsel for the biotech firm Affymetrix, told the same hearing, "There should be no patenting of gene sequences, period. They were invented by nature."

Exorbitant Licensing Fees

Because patent holders have a monopoly on certain medical tests, there is less of a free market to regulate the price of tests. In some cases these tests can save people's lives, and in other cases they are necessary for the furthering of research. The Association for Molecular Pathology said in a statement, "Increasingly, patent holders or their licensees are choosing to monopolize genetic testing by preventing all other health-care providers and facilities from performing tests covered by the patents." They warn that, "The use of patents or exorbitant licensing fees to prevent physicians and clinical laboratories from performing genetic tests also limits access to medical care, jeopardizes the quality of medical care, and raises its cost." Some in the biotech industry counter by saying that if there were no patents the expensive research to develop the tests would not have been done.

CAUTION

BioWarning

The American Medical Association, in a report on gene patenting, highlights a particularly egregious example of how it can stop vital medical testing. Families of children with a genetic disease called *Canavan disease* that primarily affects Jews of Eastern European descent had approached medical researchers and asked that they study the disease. The families agreed to be studied, and the gene causing it was found. As a result, it was recommended that couples with an Eastern European Jewish background be tested for the gene in prenatal screening. However, the Miami Children's Hospital had received a patent on the gene and as a result of the licensing agreements required for the tests, the tests were not available to those who needed them—even to the families who had been part of the study.

Red Tape Ties Up Research

Because there are so many biotech patents, researchers can get tied up in red tape when they instead need to be doing research. Not only do they need to research patents, but they also have to go through the steps of obtaining licenses and paying for them when they want to use a technology that has been patented.

For example, scientists who worked on "golden rice" which has been genetically engineered to provide Vitamin A and might be a boon to the malnourished in Third World countries, had to get permission from more than 40 patent or contract holders in order to do their work, reports *The New York Times*. As of this writing, golden rice is still not available. There is a chance that if researchers could have concentrated on perfecting the rice instead of obtaining licenses, it could have been available earlier.

The Government Paid for the Research

Some people note that many scientists responsible for genetic breakthroughs have been paid by the government in one way or another, in particular through the many federal grants through agencies such as the National Institutes of Health, the National Science Foundation and the Department of Energy. Because of that, they say, genetics knowledge should be freely available, and not owned privately. Others counter that it is a legitimate role of government to fund research that will spur growth of the private economy.

Life Shouldn't Be Patented

There are those who believe from a philosophic or religious point of view, that life should not be patented. Life, these critics say, is not something that should be able to be bought and sold like any other commodity.

BioSource

There is a movement to get Congress to pass a law outlawing patents on both life-forms and genes. For a copy, go to www.gene-watch.org/programs/patents/petition.html.

Patents Will Concentrate Power in Corporations

A few critics contend that biotech and genetic technologies will become the economic powerhouses of the twenty-first century, and that if private corporations can gain control of the use of genes, they will control most economic activity. The economist and author Jeremy Rifkin, who would ban patenting genes, told *Red Herring* magazine: "Whoever controls the genes controls the (twenty-first) century. Genes are the raw resources for everything important to commercial life in the (twenty-first) century, just as fossil fuels were to the last 200 years. And the name of the game is patents." It should be noted, however, that Rifkin has attempted to block every step of the biotech revolution, originally opposing even the simplest isolation of genes in the laboratory.

Universities Fight Back

A group of prestigious universities, backed by nonprofit foundations, has come up with a solution that they believe will help solve the patent problem. The universities claim that biotech patents held by large corporations have been stifling the development of crops that can help feed millions of hungry and malnourished people. And they further claim that biotech corporations have little interest in developing crops to help feed the developing world.

The universities are joining together and letting any patents developed at the universities be used more freely by anyone who wants them. They will also put their patents into a database so that anyone who wants can easily use them.

The effort, called Pipra, for Public Sector Intellectual Property Resource for Agriculture, is made up of Cornell, the University of California, the University of Florida, Michigan State, North Carolina State, Ohio State, Rutgers and the University of Wisconsin. It is also supported by the Rockefeller Foundation and the McKnight Foundation. Included in the effort are also two research labs, the Donald Danforth Plant Science Center in St. Louis and the Boyce Thompson Institute for Plant Research in Ithaca, New York.

BioFact

It's unclear whether generic drug makers will manufacture bioengineered medicines when those drugs' patents expire. When a drug patent expires, generic drug makers can manufacture it if they can prove that it is exactly like the original. But creating bioengineered drugs may be too costly for generic drug companies. Even tiny changes in drugs can cause different and potentially harmful reactions. *The New York Times* reports, "a bioengineered anemia medicine made under license by two companies; one version had no serious side effects while the other did. No one has figured out why."

The Least You Need to Know

- For an invention to receive a patent, it has to be "nonobvious" and useful, and the patent application has to describe the invention in a way that others can reproduce it.
- A gene can be patented only if it can be isolated outside the body, if the function of the gene is understood, and if a useful commercial purpose can be suggested.
- The first patent to a bioengineered life form was given to Ananda Chakrabarty for bacteria that could eat oil.
- Proponents of patents for bioengineered plants and animals and for genes say that it leads to innovation and new medical treatments.
- Critics of patents for genes claim that patenting stifles innovation because knowledge cannot be shared freely any longer. They also claim that it hurts medical care because it makes some medical tests prohibitively expensive.
- Some people are trying to get Congress to pass a law that would prohibit the patenting of genes.

Chapter 12

The Ultimate Mini-Me

In This Chapter

- A look at the original pro-cloner, Dr. Richard Seed
- What the UFO cult the Raelians say about cloning
- A brief review of what it means to clone humans
- Why people believe human cloning should be allowed
- An interview with prominent pro-cloner Randolfe Wicker

Ever since Eve was created from Adam's rib, humankind has been fascinated by cloning. You'll find cloning stories and myths in many religions, for example, such as cloned demons being created from the blood drops of a "blood-demon" in Hindu mythology. And there have been countless novels and movies in which cloning has played a central role, from *Brave New World* to *Boys from Brazil* in which Hitler is cloned. In *Multiplicity*, Michael Keaton is cloned because he is so overworked and overtaxed that only several versions of himself can do all that needs to be done in his overscheduled life.

Although today most people would ban human cloning, there are still many people who are in favor of it. They cite a number of benefits for individuals and society that would be brought about by human cloning. In this chapter, we'll take a look at why some people favor cloning, and we'll also examine some of the colorful personalities and some charlatans who claim to have already cloned humans.

Understanding Human Cloning

Before we discuss all the reasons people believe that human cloning should be pursued, let's review for a moment some basic facts about cloning and human cloning. (For more details, turn to Chapter 6.) Here's the basics you need to know about cloning before going into a discussion of why some people believe it's a good thing:

- You clone an animal by taking the DNA out of an egg, then taking a cell from the animal you want to clone, and merging the DNA from that cell with the egg and forcing the egg to divide. The fertilized egg forms an embryo, and the embryo is then implanted into a mother. A cloned animal is then born in a normal fashion.
- The cloned animal will have the genetic makeup of the animal that donated the cell—not the animal that donated the egg.
- Currently, cloning technology is in its infancy. Most attempts at cloning fail, and many animals that are born as a result of cloning are unhealthy.
- Even though an animal is cloned, that does not mean that it will be identical to the animal from which it was cloned. Environmental factors turn genes on and off, and the genes of a cloned animal may be expressed in different ways than in the animal from which it was cloned.
- Scientists generally agree that it is not yet possible to clone a human being. No *primates*—including humans—have yet been successfully cloned because when the DNA is taken out of the egg, important proteins that aid in cell division are also removed, and the embryo does not develop properly.
- Even though groups like the Raelians claim to have cloned human beings, there is no evidence that a human being has ever been cloned, or is close to being cloned.

> **BioDefinition**
>
> A **primate** is an order of mammals which includes apes, monkeys, humans, and "prosimians" which are small, tree-dwelling, nocturnal animals such as the tree-shrew and lemurs. Primates arose during the late Cretaceous period as forest dwellers.

- There is a huge difference between therapeutic stem cell cloning and cloning a human being. In therapeutic stem cell cloning, stem cells are taken from a cloned embryo and then used for research or medical purposes. Many people believe that stem cells can help cure many diseases. In this chapter, we look at human cloning, not stem cell cloning. For the pros and cons of therapeutic stem cell cloning, see Chapter 14.

In the Beginning Was a Seed

There have been a number of controversies over the years surrounding those who want to clone humans or those who have claimed to clone humans. In many cases, those making cloning claims have been on the fringe of science, or have been working outside their scientific specialty.

The first person to make big news in relation to human cloning was Dr. Richard Seed—a more than ironic name for someone who wants to pursue cloning. Seed generally uses the honorific Dr. in front of his name because he has a Ph.D. from Harvard in nuclear physics. He does not, however, have a medical degree or one in life sciences.

In 1998 Dr. Seed announced that he planned to set up a human cloning clinic. He also said that he planned to clone himself. In 1999 he amended that to say that he would clone his wife Gloria, and that Gloria would also have the cloned embryo implanted in her so that she would bear her own cloned child as well.

> **BioFact**
>
> In 1998 Dr. Seed predicted that he would clone a human being in a year and a half to two years, which means that if he met his goal, a human would have been cloned by 2000. His predication was far off the mark—a human being has not yet been cloned, and it's unclear when one might be cloned.

Dr. Seed said that he planned to clone humans to help infertile couples have children who would be genetically related to one of them. He also said that cloning could be used to replace a family member who died or was killed. And he said that the very act of researching human cloning would result in other medical breakthroughs that would help humankind. Seed also had something of the mystic about him and claimed that cloning is "the first step in becoming one with God."

His plans were to set up a cloning clinic in Chicago, and then follow that up with similar clinics in 10 to 20 other locations in the United States and five or six internationally. He said that once he perfected his procedure, 200,000 humans would be cloned a year.

None of this came to pass, of course, and Seed hasn't been heard from in years.

What Do UFOs Have to Do with Cloning?

Seed may have appeared to have a marginal knowledge of cloning, since he was not a medical doctor and had no degree in biology. But he seems positively overqualified when compared to the Raelians, who burst into public consciousness in 2002 when they claimed to have cloned a human being called baby Eve.

They claimed that the baby had been cloned through their company Clonaid. But they never produced the baby or the mother, or revealed the scientific details of how the cloning was allegedly done.

The Raelians are headed by a French former racing car journalist who calls himself Rael (his birthname was Claude Vorilhon). Rael claims that human beings were created in laboratories by aliens and then brought to our planet. The aliens are called Elohim which in ancient Hebrew means "those who came from the sky."

BioFact

Judging by his beliefs, Rael certainly has a vivid imagination. But when it comes to his descriptions of the aliens who revealed the Raelian religion to him, he falls back on the oldest of clichés. Rael says that the aliens are a little over three feet tall, have almond-shaped eyes, and have pale green skin—the very picture of a prototypical little green Martians. The only way they differ from the cliché is that they have long dark hair instead of being bald.

It appears that the cloning claim is somehow bound up in the Raelian's religious beliefs. They believe that human beings have souls, but those souls die when the body dies. However, if a person's DNA can be recreated, then the soul will live again. So constant cloning would be the key to eternal life.

In fact, the goal of Raelian cloning appears to be eternal life for its members, by cloning them and then somehow transferring the soul, personality, and memory into that newly cloned being. According to the group's Clonaid.com website:

> Once we can clone exact replicas of ourselves, the next step will be to transfer our memories and personality into our newly cloned brains, which will allow us to truly live forever. Since we will be able to remember all our past, we will be able to accumulate knowledge ad infinitum.

Among the Raelian's many beliefs are that the Elohim cloned Jesus twice—first when he was born (which would explain Mary's virgin birth), and then again after he died (which would explain his resurrection).

Like Dr. Seed, the Raelians have largely dropped out of sight, although they are still active and trying to recruit members.

BioSource

The website of the Raelian's company Clonaid can be found at www.clonaid.com. At the writing of this book, the site claimed that the company had cloned five babies, although they offered no proof. The site hawks a variety of services and products, including its "Insuraclone" service which promises to preserve your cells and genetic material forever, in case you ever want to clone them, for $200 a year plus an undisclosed initial fee that varies by person. There is also a "genetic repair kit" it will sell for an undisclosed fee, as well as a Clonapet service for cloning your dead pet (no fee given as well). Then there's the "embryonic cell fusion system" called the RMX2010 that the company will sell to you for $9,220—excluding shipping and handling charges, of course. And if you want to buy or sell human eggs, the company claims it will do that as well.

Reasons for Cloning

Dr. Seed and the Raelians are not the only people who want to clone human beings—far from it. And the truth is, most people who favor the cloning of humans believe cloning should be allowed for legitimate reasons. They are not wild-eyed fanatics, not people looking for the media spotlight, but rather people who have given a good deal of thought to the issue and have a coherent set of reasons why they believe cloning would help individuals and mankind.

They recognize that the technology does not currently exist to allow for the cloning of humans. So when they speak in favor of cloning, they are not talking in terms of today or next month. Rather, they're talking long-term. And they are speaking of allowing cloning after it is no longer a dangerous procedure. So when we cover the reasons they believe cloning should be allowed, assume that they're asking that humans be allowed to be cloned only after it is no longer a dangerous procedure to the mother and child.

With that said, let's look at the main reasons why people believe cloning can be a good thing.

To Help Infertile Couples

One of the primary reasons put forward by those who favor cloning is that it can help infertile couples have children who are genetically related to one of them. If a man,

> **BioSource**
>
> The most comprehensive source of pro-cloning related information online is run by the Human Cloning Foundation at www.humancloning.org. You'll find a great deal of pro-cloning related material and resources there.

for example, could not produce sperm, he could still produce offspring by cloning. And if a woman could not produce eggs, she could produce offspring by cloning as well.

Of course, the resultant clone would be genetically related to only one of the parents. However, parents could conceivably clone more than one child—one of the father and one of the mother—and so could have a family that was a mix of their genes.

Those who favor cloning say that doing this is really only an outgrowth of current reproductive technologies to help infertile couples, such as in vitro fertilization. With in vitro fertilization, eggs from the mother and sperm from the father are mixed together in a laboratory and a resulting embryo is then implanted in the mother. The only difference between that and cloning, say those who favor cloning, is that instead of an egg being mixed with a sperm, an egg minus its DNA is mixed with another cell to form an embryo. The embryo in both cases is then implanted in the mother.

Cloning Is a Reproductive Right

Many people in favor of cloning say that it should be allowed because it is a reproductive right. They are not necessarily saying that people *should* clone themselves; instead, they say that people have a *right* to clone themselves, in the same way that they have a right to other reproductive-related technologies and procedures such as contraceptives or in vitro fertilization.

They believe that the government should not be allowed to tell people what they can or can't do when it comes to reproduction.

In this light, it would also give homosexuals the right to have children related to themselves. In a lesbian couple, for example, one woman could be cloned and then the embryo brought to term in either of the women. In a male homosexual couple, one man could be cloned, but they would have to find a woman to donate an egg and to agree to bring the embryo to term.

To Clone for a Cure

Imagine that a child had an incurable disease, and the only way to cure him in later life would be with a transplant of some kind—a kidney or a bone marrow transplant. Also imagine that no donor with a match for the organ can be found, and so the chances that the child would reject any transplant would be high. Without a transplant, the child will die.

Those who favor cloning say that the child could be cloned, and then in a few years, the resulting clone could donate an organ or bone marrow. The older child would be saved, and the resulting younger cloned child would live as well, because bone marrow regenerates, and people can live with only one kidney. In that way, parents would have two healthy children, but without cloning, their only child would have died. So they would have a new life—the new child—and save an existing life.

BioFact

Cloning a child to help another sick child is not far-fetched—a related case has already happened. A couple in Great Britain had a child born with Diamond Blackfan anaemia, a genetic blood disorder that doctors hoped to cure by implanting stem cells from the umbilical cord of a baby who is an exact tissue match for him. The couple wanted to undergo in vitro fertilization, but wanted to make sure that the embryo they chose would be a tissue match for the child before implanting the embryo in the mother. Because the British government body Human Fertilisation and Embryology Authority (HFEA), which regulates fertility clinics, refused to allow the couple to undergo the therapy, they came to the United States to have it done.

To Have Children Free of Genetic Disease

Cloning can help couples ensure that they would have offspring free of deadly genetic diseases. For example, if both parents had a recessive gene for a deadly genetic disease, for any child of theirs there would be a one-in-four chance that the child would inherit that deadly disease.

The genetic disease only shows up if a child has two genes for it, one from each parent. Because each parent has only one gene for the disease, each parent is free of that disease. So if a parent is cloned, the resulting child will not have the genetic disease, because he will have the same genetic makeup of one parent—only one gene for the disease.

BioFact

For more information about genetics, heredity, and dominant and recessive genes, see Chapter 2.

To Recreate a Lost Child or Relative

Perhaps the worst tragedy that can befall a parent is the loss of a child. Cloning, say its adherents, would allow a parent to recreate a dead child or relative. In this instance, when a child is dying, cells would be taken from the child so that the child could later be cloned. While the new child would not actually *be* the child who died, cloning adherents say, it would in some way help take the sting out of the death, by being a new child who looked very much like the deceased one, and had the same genetic makeup.

The Report of the President's Council on Bioethics cites a similar example to this as a reason why someone might want to clone a child. In an imaginary case in which a father, mother, and child were involved in a deadly car accident, the father is killed instantly, and the child is critically injured and dying. If the mother took cells from the dying child, and then cloned that child, it could "allow her to preserve a connection with both her dead husband and her dying child, to create new life as a partial human answer to the grievous misfortune of her child's untimely death, and to continue the name and biological lineage of her deceased husband."

BioFact

The Raelians have made a number of outrageous cloning claims. Among those claims is that they were able to retrieve living cells from the bone marrow of a dead child that had been buried for four months, and that they were in the process of cloning a new child from the dead child's cells.

The President's council came out very strongly against human cloning, and so did not consider this argument particularly compelling. It cites a wide variety of reasons why it believes that human cloning should be banned. For more information, see Chapter 13.

To Make up for a Bad Childhood or Life

The pro-cloning site www.humancloning.org cites a number of reasons for cloning children, including a very surprising one—that many people "would like to have a clone so that it may lead the life that was meant to be theirs." According to the site, many people have written to it to say that they want to have a clone because they have had a very difficult life, and they hope that their clone would be able to have an easier one. Because of their difficult life, the site says, they "feel robbed of the opportunities they should have had in life." So these people would like to give the clone the opportunities that they never had in life. According to the site, "Some see this life as a sacrifice so that the life of their clone may be enriched."

To Live Through a "Later-Born Twin"

Some pro-cloners have taken to calling human clones "later-born twins" of the person whose genetic material was cloned. The idea is that since the cloned being has the same genetic material as the original, they are twins. And it is a "later-born twin" because it is born decades after the person whose genetic material is being cloned.

Some people feel they would have an immortality of a kind in having this "later-born twin" because their DNA would live on, even when they die. And there are those who want to have a "later-born twin" because they believe that twins have a special relationship that other people don't have, and they want to have that same sense of closeness that twins have. In this way of thinking, an especially strong bond would be formed because the resulting child would be both a child and a "later born twin" to the parent.

BioFact

The U.S. public does not want there to be research into human cloning, but favors research into therapeutic stem cell cloning. That's the finding of a nationwide survey in 2002 conducted by the Pew Research Center and the Pew Forum on Religion and Public Life. The survey found that people oppose research into human cloning by a margin of 77 to 17 percent, but favor government funding for stem cell research by a margin of 50 to 35 percent.

A Talk with Randolfe Wicker

As you've seen in this chapter, the human cloning movement has had its share of charlatans. But not everyone who favors cloning is a charlatan—in fact, far from it. There are many well-meaning people who favor cloning as well.

One of the most prominent people pro-cloning activists is Randolfe Wicker, a New York man in his 60s, who says that he was a pioneer in the gay rights movement and fighting for civil rights for gays. He also claims to have led the first pro-cloning demonstration. Wicker is the founder of the Clone Rights United Front, spokesperson for the Reproductive Cloning Network, and former CEO of the Human Cloning Foundation.

Wicker has appeared on TV and radio explaining why he believes cloning should be allowed, and has led a number of public demonstrations in favor of cloning, including one in which he handed out buttons saying "Clone Jesus." He is witty, entertaining, outrageous, has something of a scattershot personality, and he signs his letters and emails with the signoff, "Cloningly Yours."

We interviewed Mr. Wicker several times for this book, and in this section we'll give a brief summation of his views on human cloning.

The overall reasons he favors cloning, he told us in an e-mail:

> Human cloning enhances life and the dignity of human beings. It makes every person fertile. It enables every human being to see his or her genotype live on into another lifetime. It enables humankind to preserve those rare geniuses that are born only now and then—Einstein, Thomas Edison, Mozart, and so on.

BioFact

Wicker has testified before Congress to spell out the reasons that he favors cloning. To read his testimony, go to www.clonerights.com/official_documents.htm.

Wicker says that he has seen frequent disharmony in families, and believes that if parents were to have "later-born twins," there would be more harmonious family relations because of the special feelings that people have for their twins. He believes that many people want to have their genes live on through "later-born twins" and that they have a right to do that.

Wicker holds that cloning is, "an individual reproductive choice," and the government has no right to ban it, in the same way that it should not be able to ban any other kind of reproductive technologies.

He concedes that most people would ban cloning, but he believes that ultimately they may come around to his point of view. He held a demonstration once where he said that Princess Diana should have been cloned, and says that the reaction to the demonstration surprised him—that many people were in favor. He concludes, "The reality is that while people are intellectually against cloning, they are not against cloning someone they know or admire."

Wicker believes that human cloning research will lead to medical breakthroughs, and points out that it can be used to help save the lives of children, by cloning sick children, and then using the cloned child to donate bone marrow to the sick child, for example.

And in the long run, he says, "at its very heart, cloning has elements of religious belief in it. It's hard to understand why, but even my own feelings, my own desire to be cloned shows it. These feelings are so strong, they have some kind of underlying value basis. Why else would religious groups feel so strongly about it?"

Although Wicker would want to be cloned, he says he will not do it, because he believes he is too old to have a child. He claims that an Indian scientist asked if he would allow himself to be cloned, and he turned the scientist down.

Wicker believes that even with all the current risks to human cloning, it should be legal today and allowed today.

> A couple has the reproductive right to say they want to have children that are later-born twins of themselves … They have the right to take a chance, as long as they know the risks and know that they are going into uncharted territory. You have to have people take risks and take chances and go where no one has gone before for human progress to take place.

Ultimately, he believes, human cloning will be legal and accepted. "The technology will advance," he maintains. "Human cloning is inevitable, and at some point society will have to deal with it in an informed and rational way."

The Least You Need to Know

- Dr. Richard Seed said in 1998 that he would clone a human within two years, and would have two dozen cloning clinics. He never cloned a human and no clinics ever opened.
- Even though the Raelians and some others claim to have cloned humans, there is no evidence that a human has ever been cloned.
- Those who favor cloning say that cloning is a basic reproductive right that should not be banned by an intrusive government.
- Pro-cloners say that cloning can help infertile couples have children, and can help parents with recessive genetic mutations make sure that their children are born without genetic disease.
- Those who favor cloning say that a sick child could be cloned, and the resulting clone could donate bone marrow or a kidney to the sick child—and then the sick child and the clone could lead healthy lives.
- Pro-cloners say that cloning can recreate a deceased child or relative and so help a family through grief while creating new human life.

Chapter 13

The Case Against Human Cloning

In This Chapter

- How cloning can harm a child's sense of self
- The confusing family relationships that cloning would bring
- Why cloning could psychologically harm the cloned child
- The medical dangers posed by human cloning
- How human cloning could harm society

As we discussed in Chapter 9, human cloning is not possible today, despite the claims of some groups and scientists that they have already cloned a human being.

But science has a way of making today's impossibility into tomorrow's reality. So, the human cloning debate rages on. In this chapter, we'll look at the many reasons why many people and groups want to ban human cloning. Note that we don't cover the debate over therapeutic stem cell cloning here. For that, turn to Chapter 14.

Why Should We Care About Human Cloning?

The United States is a freewheeling kind of place with a powerful libertarian streak and places a premium on human freedom. Unless there is a compelling reason to place limits on what we can do, we like the government to stay out of our business.

Even though the country has a Puritanical heritage, and despite a powerful religious political voice, this libertarian streak is the reigning ethos of our country.

So why are there so many calls to ban human cloning? Why does this issue seem to unite the left as well as the right, and atheists and the religious alike?

In short, why should we care? Apart from feelings of personal repugnance toward the technology, why should society play a role in cloning at all? Shouldn't it be left up to the individual to decide?

As you'll see in this chapter, there are a variety of medical, societal and religious reasons why many people would ban cloning. But the question remains, in a generally libertarian society, why should society care about cloning? Isn't it a matter of personal preference?

The answer has to do to a great extent with what kind of society we might become if human cloning were allowed. It's not merely the rights and medical condition of the cloned human being that we are concerned with—it is what kind of culture and society human cloning would lead to.

BioFact

At the writing of this book, there was no law that banned human cloning in the United States. (That may have changed by the time you read this.) The federal government banned the use of any federal research money to study the cloning of human beings, but that doesn't mean that privately funded groups could not try to clone human beings. The U.S. Congress has considered laws to ban human cloning, but politics keeps cropping up: Republicans have written the law to ban therapeutic stem cell cloning as well as human cloning, while Democrats and moderate Republicans want to allow research on therapeutic stem cell cloning. Because the Republicans won't separate the two issues, no law has been passed as of yet.

That's why you'll find that some of the loudest voices raised in the debate are those who normally take a laissez-faire attitude when it comes to placing any limits on human freedom.

With that as a background, let's start off by taking a look at an issue that sometimes gets lost in the debate—the implications of cloning for the being that gets cloned.

Protecting the Rights of Children

One of the most powerful arguments against cloning is how it will harm the child who will be born as a result of the procedure, not only medically but in a variety of other ways as well. We'll explore each in this section. In the next section, we'll look at the medical problems with cloning.

Destroying a Sense of Self

One of the most difficult, and the most satisfying things about being human is developing a sense of self—understanding our particular capabilities, wants, needs, strengths, and understanding how we fit into the world.

A vital part of this is learning from and breaking away from our parents and in understanding how we are similar to and different from our parents.

But if someone is a mere clone of a parent, what sense of self can there be? Will he feel he is fated to merely be an imitation of a parent, and so has no free will? Will he think that his life is already predetermined for him?

CAUTION

BioWarning

Those who fear that human cloning technologies may be used to bioengineer genetic freaks most likely felt a shudder of apprehension when they heard the news that scientists at the Foundation for Reproductive Medicine in Chicago had created a so-called "she-male" human embryo. The scientists injected cells with male DNA (they had the Y chromosome) into female embryos. The embryos were not allowed to develop past six days.

Additionally, most people feel the weight of expectations from their parents—how heavy will that weight be knowing that a parent wanted an exact replica of himself?

Worse still, what of those people who were cloned from the cells of a dead child—imagine the expectations and sense of guilt that would accompany that person through life.

Furthermore, many people believe that our very birthright is a sense of uniqueness and possibility, and so by taking it away you are taking away a person's very birthright.

In fact, the bioethicist Evelyne Shuster of the University of Pennsylvania argued in a paper prepared for the International Symposium on Bioethics and the Rights of the Child that for this reason, "Cloning violates the child's right to an open future. This is a cruel, inhuman and degrading treatment of children that violates the Universal Declaration of Human Rights."

Confusing Family Relationship

There's an odd old song titled "I'm My Own Grandpa" in which through a variety of odd intermarriages, someone ends up as his own grandfather.

That's something like the cloned would face, except worse. The confusing family relationships they would have to endure would be comical if it weren't so humanly damaging. Consider the relationships of someone who was a clone of his father, and who came to term in his mother:

- He would be a brother to his father, essentially, because he would have the same genetic material as his father—he would essentially his twin.
- He would be his mother's brother-in-law, because he would be a brother to his father.
- If he had any siblings who were born to his father by natural means, he would not only be their sibling, but also their uncle because he is a twin brother to his father, and therefore their uncle.
- If he had any children, he would also be their great uncle—they are his father's grandchildren, but because he is also his father's brother, they would be his grandnieces and grandnephews.

BioFact

Want to find out what odd series of marriages leads to someone being their own grandfather? Go to users.cis.net/sammy/grandpa.htm to read the lyrics to "I'm my Own Grandpa." There's even a musical accompaniment, so you can sing along.

If that's not confusing enough, what happens if his mother now had a clone? And if his mother had a clone, couldn't he marry the clone, because they would not be genetically related? In that case, he would be his sister's husband, his mother's son and son-in-law, and his father's son, brother and son-in-law. The children he has with his sister would be … but let's not go there, because there may not even be words to describe the family relationships that would result. You get the point though. Familial relationships help define who we are as individuals, and to a great extent are the bedrock upon which society is based. Cloning would cloud and destroy any sense of those relationships.

Cloning Turns Children into Commodities

Producing a child by cloning turns that child into a commodity—in return for payment (and a very high one, no doubt), an offspring is provided with a specific genetic makeup. It is no different than buying any other commodity whose attributes you know ahead of time. Conceivably, those cloned commodities could then be sold to the highest bidder. For example, if parents wanted to have a child with scientific potential, would they be willing to pay top dollar for the cloned embryo of a Nobel Prize winner? Would parents hoping for a big payoff for a child with professional athletic ability pay top dollar for the cloned embryo of a top professional athlete?

Parents of Cloned Children Will Be Unsuitable

An argument can be made that anyone who would to clone a child is, by nature, an unfit parent. It's not merely the parents' egoism that would make them unfit—it's that they would also subject their own child to the various psychological problems and medical dangers that necessarily attend cloning. Anyone who cares so little about the fate of their child would be unfit to parent.

BioFact

In one sense, there is no such thing as a true clone. Even if two beings have the same genetic material, it doesn't mean that those genes will be expressed in the same way. Environmental factors play a role in which genes turn on and off, and so a clone may well look and act differently than its parent. There are also biochemical reasons caused by the cloning process that will cause the DNA to express itself differently. Because of this, nuclei donated from two different types of cells from the exact same donor could give somewhat different features to clones even though they contain the very same DNA sequence.

A Cloned Child Cannot Give Consent

Cloning has severe potential psychological and medical problems associated with it. Standard medical practice requires that if someone is going to be subject to an experiment that carries dangers, they be informed about those dangers, and then they can decide whether to consent to the procedure.

However, an unborn child has no means of giving consent. That means that child will be subject to experimentation without his will or knowledge.

Medical Dangers of Human Cloning

As we explained briefly in Chapter 6, there are a variety of medical dangers associated with cloning animals. Clones, as a general rule, are unhealthy animals and there are reasons to believe that any human clones may be even unhealthier than the clones of other mammals.

For more details about the general dangers of cloning, turn to Chapter 6. Here's a brief synopsis of potential dangers in mammal cloning:

- Cloned mammals die younger than non-cloned mammals and suffer prematurely from diseases of old age, such as arthritis. For example, Dolly the sheep died young and suffered from arthritis and other maladies.
- Cloned animals are at a higher risk of having genetic defects and of being born diseased and deformed.
- Studies have found that cloned mice have died prematurely from tumors, damaged livers, and pneumonia.

BioFact

There is a big difference between human cloning and therapeutic stem cell cloning. In human cloning, a human being would be cloned by implanting the DNA from the person to be cloned into an egg cell whose DNA has been removed. The resulting embryo would then be brought to term in a mother's womb and the end result would be a cloned human being. In therapeutic stem cell cloning, special cells called stem cells are taken from an embryo of less than a few days old and used in research, and potentially to help cure many diseases, including Parkinson's disease, Alzheimer's disease, and countless others.

- The Whitehead Institute for Biomedical Research in Cambridge, Massachusetts studied 10,000 genes in the livers and placentas of cloned mice and found that hundreds of the genes had abnormal activity patterns.
- Some cloned animals appear healthy at birth, but die for no known reason.
- Even clones that appear normal have subtle differences in their genetic makeup when compared to the animal from which they were cloned, concluded

Rudolph Jaenisch of the Whitehead Institute. Something appears to go wrong in the process that switches genes on and off. "This suggests that even apparently normal clones may have subtle aberrations of gene expression that are not easily detected in the animal clone," he stated. Because of this, he called attempts to clone humans "dangerous and irresponsible."

BioSource

The scientist who cloned Dolly the Sheep, Iam Wilmut, has taken a strong stand against human cloning. With Rudolph Jaenisch of the Whitehead Institute, Wilmut wrote an editorial in the influential journal Science warning of the dangers of cloning and recommending that there be a moratorium on human cloning.

Dangers to the Mother

It's not only the child who will be facing medical dangers; the mother will as well. The Report of the President's Council on Bioethics chaired by Dr. Leon Kass, warns that animal studies "suggest that late-term fetal losses and spontaneous abortions occur substantially more often with cloned fetuses than in natural pregnancies. In humans, such late-term fetal losses may lead to substantially increased maternal morbidity and mortality."

The report also notes that pregnancies using cloned fetuses often lead to serious medical consequences, and cites a study showing that nearly one-third of cows pregnant with cloned fetuses died from late-pregnancy complications.

Because of this, the National Academy of Sciences concludes, "Results of animals studies suggest that reproductive cloning of humans would similarly pose a high risk to the health of both fetus or infant and lead to associated psychological risks for the mother as a consequence of late spontaneous abortions or the birth of a stillborn child or a child with severe health problems."

BioSource

You can read the entire report of The Report of the President's Council on Bioethics, "Human Cloning and Human Dignity: An Ethical Inquiry" at www. bioethics.gov/reports/ cloningreport/fullreport.html. If you prefer, you can also buy a copy at a bookstore or online. It has a retail price of $14. For more information about the council, go to www. bioethics.org.

It's Even Worse for Humans

There is evidence that cloning may be even more dangerous in humans than in other mammals. For reasons explained in Chapter 9, primates including man have not yet been cloned because the cloning process removes proteins that are necessary for cells to divide normally and pass along genetic material properly. Primate embryos show an extremely high degree of malformation and genetic problems. In fact, cloned primate embryos have been called a genetic "gallery of horrors" by a scientist involved in cloning primates. Some cells contained no chromosomes, some contained double the number, some had strange combinations of chromosomes, and they were often spread throughout the cells rather than in a well-defined nucleus.

Ian Wilmut, the scientist who cloned Dolly the Sheep, told the BBC that, "The most likely outcome of any attempt to do that (clone a human) would include late abortions, birth of children who would die, and worst of all the birth of children who would survive but would be abnormal."

Clones Would Be Experiments

By the very nature of the technology, human cloning is experimentation. While it may be theoretically possible that some day human cloning could become a relatively simple medical procedure, most people don't believe that will ever be the case. And even if it were, the first several generations of cloned children would be experiments, because it will take a great deal of experimentation before the procedure could be perfected. That means that all those children—and the women who bring them to term in their wombs—will be little more than research subjects.

BioFact

The National Academy of Sciences recommended in 2002 that cloning should be banned for at least five years in the United States. The academy based its findings solely on the medical dangers of cloning, and didn't consider ethical or religious issues. The Academy concluded: "Human reproductive cloning should not now be practiced. It is dangerous and likely to fail. The panel therefore unanimously supports the proposals that there should be a legally enforceable ban on the practice of human reproductive cloning." However, the Academy also recommended that therapeutic stem cell cloning should continue because of its many potential benefits. For the National Academy's report on the issue of human cloning, go to www.nap.edu/books/0309076374/html/.

Societal Dangers of Cloning

It's not merely individuals who would be harmed by human cloning, say the procedure's many critics. Society as a whole would be harmed as well to such a degree that it would not be the kind of culture that many of us would want to live in. In fact, the

"Report of the President's Council on Bioethics" warns, "The impact of human cloning on society at large may be the least appreciated, but among the most important, factors to consider in contemplating the morality of this activity."

We'll cover the various societal dangers that cloning poses in this section.

Cloning and Eugenics

In the early decades of the twentieth century, a pseudoscience called "eugenics" became popular. It was an attempt to breed a "better" human race by encouraging those with "good" genes to have children, and discouraging those with "bad" genes from having children. The natural outcome of this movement was the Nazis, with their attempted extermination of the Jews and others they deemed undesirable, and their attempt to breed a master race.

BioSource

Eugenics was not a cult movement, or one that was outside the mainstream of polite society or politics. In fact, it was a powerful force at the center of American political life. In the United States it managed to get passed a variety of laws that forcibly kept racial and ethnic groups separate from one another, that restricted immigration from southern and eastern Europe, and that even sterilized "genetically unfit" people so that they could not have children. For a history of the movement, including online virtual exhibits, go to the Eugenics Archive run by the Cold Spring Harbor Laboratory at www.eugenicsarchive.org.

Cloning could well lead to a new, even more effective form of eugenics. In countries run by dictatorships, the government could sponsor a mass cloning campaign and clone people who it deemed had the proper genetic makeup. This could lead to a far more effective form of eugenics than even the Nazis dreamed of.

But even in democratic societies in which the government might not be involved in cloning, cloning could lead to a kind of free-market eugenics, particularly when it is combined with bioengineering techniques. For example, an adult may have always wished that he could have been born with blue eyes instead of brown eyes, and blonde hair instead of brown hair. In theory, he could clone himself and bioengineer his clone to have blue eyes and blonde hair through manipulating the genes to be placed into the cloned embryo.

On a mass scale, this could lead to a kind of master race being created, one based on fashion, perhaps, or on what attributes people think will lead to a cloned offspring's financial benefits. Society would then not be populated by the diverse, energizing mix of people and their differing features and unique talents that we have today, but instead increasingly by a homogenized "ideal" human. And theoretically that human ideal could be one defined by advertising executives—imagine a world in which biotech companies vied with each other to sell the "best" or most "desirable" genetic material to those who want to be cloned, and so flood the media with advertising to convince people that their cloning procedures and super-value genes are the ideal ones. Far-fetched, yes. Impossible? Perhaps not.

CAUTION

BioWarning

If you'd like to read a novel warning about the dangers of government-sponsored mass cloning, turn to the classic *Brave New World* by Aldous Huxley. This chilling and prophetic novel was published in 1932, well before any animal had yet been cloned, and decades before the discovery of the physical structure of DNA. Yet it describes a world in which mass cloning through a technique called Bokanovsky Process populates a world with people bred to have specific sets of genes—and different genes are used for different societal castes. For a more contemporary novel that looks at the results on society not just of cloning, but of widespread bioengineering, read Margaret Atwood's novel *Oryx and Crake*.

It Changes Society's View of Children

To a great extent, a society revolves around the way it views and treats its children. When we think of London in the time of Dickens, for example, we often imagine it a cold, Darwinian place because of the way children of the poor were treated, as discardable, or miniature workhorses. Picture any society of which you would want to be a member, and most likely a part of that picture is healthy, happy, cared-for children.

A society that allows cloning, some people believe, would be one based on selfishness and callousness. Not only would children suffer, but society as a whole would suffer.

The "Report of the President's Council on Bioethics" warns:

> We are already liable to regard children largely as vehicles for our own fulfillment and ambitions. The impulse to create 'designer children' is present today—as temptation and social practice. The notion of life as a gift, mysterious and limited, is under siege. Cloning-to-produce-children would carry these tendencies and temptations to an extreme expression. It advances the notion that the child is but an object of our sovereign mastery.

And a society that views children in that way, it warns, is not a society in which most of us would want to live.

The Least You Need to Know

- Cloning could destroy the sense of "self" of the child who was born as a result of cloning.
- Family relationships would be forever altered by cloning.
- Cloning turns children into commodities, and conceivably cloned embryos could be bought and sold on the open market.
- Children who were cloned would most likely suffer from many debilitating diseases, including genetic defects, premature old age, and cancers.
- A mother of cloned children could face severe health problems.
- If society allows cloning, eugenics (the breeding of a master race) may become popular. Additionally, a society in which children are treated as commodities would be one in which many people would not want to live.

Chapter 14

The Debate About Therapeutic Stem Cell Cloning

In This Chapter

- What is therapeutic stem cell cloning?
- The benefits of therapeutic stem cell cloning
- Why proponents say the technique may be a miracle cure
- Why opponents favor banning therapeutic stem cell cloning
- What the federal government has done about therapeutic stem cell cloning

Few technologies are as misunderstood as therapeutic stem cell cloning—so much so that a number of people have suggested that the word "cloning" shouldn't be used when referring to it. No fully-grown animal results from the procedure, they point out, and the term makes people think that is somehow the same as cloning an animal or a human being.

In this chapter, we're going to explore the controversy around therapeutic stem cell cloning. It starts off with a refresher course in what the technique is and how it's done. (If you want more details, turn to Chapter 8.) But after that brief refresher course, we'll look at the controversy around the technology, which has become one of the most bitter ones in all of science because the stakes are so high. Proponents say that therapeutic stem cell cloning can ultimately save millions of lives, while opponents, who are primarily anti-abortion activists, claim that the research kills embryos, which they consider to be children.

Understanding Stem Cells

You have all kinds of cells in your body—stomach cells, brain cells, blood cells, kidney cells, and many others. Each of those cells is highly specialized to perform specific functions. Some stomach cells, for example, secrete enzymes that help you digest your food—but they can't transport oxygen from your lungs to the rest of your body because they're not specialized to do that.

All of these cells didn't start out specialized like this, though. At one point, you were made up of only a single cell—a fertilized egg. There were no eyeball cells or liver cells or lung cells present. There was just that single cell, an egg that had been fertilized by a sperm.

> **BioDefinition**
>
> An **embryo** is the earliest stage of the development of a plant or animal, and refers to the stage of development of a plant or animal after fertilization has taken place and before the organs have developed.

That egg quickly began dividing, but at first it divided into cells that were undifferentiated—they had yet to turn into specialized cells. You were just a tiny clump of fast-dividing cells at this point, an *embryo*. No eyes, no brain, no toes … just a few embryonic cells dividing like crazy. But each of those embryonic cells had the ability to turn into any kind of cell that your body would ultimately need, such as brain cells, eyeball cells, and so on.

These cells, which can develop into any other cell in the body, are called stem cells. Because they are from the embryo, they are called embryonic stem cells. (As we'll see later in this chapter, there are adult stem cells as well.) Embryonic stem cells have the amazing ability to turn into any type of cell in the body. They remain as stem cells until they receive a signal to differentiate into another kind of cell. They also divide very quickly when compared to differentiated cells.

Because stem cells can turn into any cell in the body and because they divide so rapidly, they represent a kind of Holy Grail for medical researchers and doctors. Since they can develop into any type of specialized cell researchers expect that they can replace damaged or diseased tissues and cure a wide variety of diseases. In fact, some believe they may be a miracle cure of sorts. (Later on in this chapter we'll look at some of their benefits.)

The Different Kinds of Stem Cells

We've covered one kind of stem cell so far, embryonic stem cells, which can turn into any cell type in the body and divide quickly. But in fact there are two different kinds of stem cells—those from embryos, and those from adults. They are related, but there are some major differences between them. As we'll see, because of these differences embryonic stem cells have a much greater promise for curing disease and for medical research.

Embryonic Stem Cells

Embryonic stem cells, as we've explained, are undifferentiated, fast-dividing cells from the embryo that can develop into any cell from the body. They can be obtained in a number of different ways, but primarily are retrieved from embryos that would otherwise be discarded in fertility clinics. Embryonic stem cells are referred to as pluripotent because they can develop into any cell in the body.

BioFact

Researchers have discovered that stem cells may be present in the amniotic fluid that surrounds babies in the womb. A team at the University of Vienna found that some cells in the fluid can create a protein called Oct-4, which is a sign that the cells may be stem cells. If true, this could be a major new source of stem cells for research and therapy, and could cool the controversy surrounding stem cell cloning. Those who oppose stem cell cloning do so because they say it is unethical to use cells from an embryo, and so using cells from amniotic fluid may not be objectionable to them. However, it is still unclear whether the amniotic stem cells have as much potential to differentiate as do embryonic stem cells.

Adult Stem Cells

Some adult tissues, such as the bone marrow, brain, blood, cornea and retina of the eye, liver, skin, and muscles have stem cells in them. These adult stem cells can build replacement specialized cells for the tissues in which they are found. If those tissues are damaged, diseased, or lost by normal wear and tear, stem cells can generate new specialized cells to replace the tissue.

Adult stem cells can only normally differentiate into the type of tissue where they are found. Stem cells in the liver, for example, cannot normally develop into retina cells, and stem cells in the cornea cannot normally develop into brain cells.

Adult stem cells are usually unipotent, which means that under normal conditions they can only give rise to one type of cell. Adult stem cells are also rare. They are difficult to identify and grow in the laboratory. And unlike embryonic stem cells, they do not replicate themselves indefinitely when grown in culture in the laboratory.

For all these reasons, they do not hold nearly as much promise for therapy and research as do embryonic stem cells. However, some experiments have shown that adult stem cells can, under certain conditions, differentiate into tissues other than the ones in which they are found. Recently, scientists introduced a survival gene into a particularly promising type of adult stem cell line isolated from bone marrow. When injected into the hearts of mice that had suffered heart attacks the new cells were able to repair some of the damaged heart tissue.

BioFact

Scientists who want to do embryonic stem cell research say that a primary source of embryos would be fertility clinics. Frequently, more embryos are produced than can be used in fertility clinics and sometimes those embryos are discarded. Stem cell researchers say that those embryos which would otherwise be discarded can ultimately save countless lives. However, there is a ban on using federal funds to do research using embryos in this manner, and so many U.S. scientists cannot use embryos from fertility clinics this way.

The Benefits of Stem Cells

Chapter 8 goes into a great detail about the benefits of stem cells, so we won't cover all of their benefits here. But they are seen as a kind of potential miracle cure by some. They may help cure many diseases such as cancer and birth defects that are caused when cells divide and differentiate abnormally. Stem cells could potentially be transplanted into brain and nerve tissue to help cure disease and conditions such as strokes, spinal cord injuries, and degenerative brain and nerve conditions such as Parkinson's disease. They could possibly treat diabetes, replace skin, clone organs, cure muscular dystrophy, and treat many, many other diseases.

Why Therapeutic Stem Cell Cloning Is Important

Those who favor therapeutic stem cell cloning do so for an obvious reason: It can lead to dramatic breakthroughs in science and medicine and may save countless lives and improve the quality of life of countless lives as well. (See Chapter 8 for more information.)

In the rest of this section we'll look at the various important groups and individuals who are in favor of therapeutic stem cell cloning.

BioFact

A potential source of stem cells in the future is so-called baby teeth, also called milk teeth. These teeth start appearing when a child is about six months old, and usually fall out when a child is between six and thirteen years old, to be replaced by adult teeth. Researchers have found stem cells in the pulp found in the middle of the teeth. Those stem cells can differentiate into tooth cells and neural cells, but no other kinds of cells.

The Medical Community Favors Stem Cell Research

The medical community rarely unites over any issue, but when it comes to stem cell research, there is a surprising agreement about the importance of the research and that it should continue.

For example, the American Medical Association (AMA) is a generally conservative organization, yet its ethics panel has come out in support of therapeutic stem cell cloning. In fact, the panel's recommendation was accepted without a single voice of dissent when it was brought to the floor of the AMA's House of Delegates Annual Meeting. The panel's report noted that stem cells could come from embryos that would otherwise be discarded for couples undergoing in vitro fertilization when they no longer needed those embryos.

The prestigious *New England Journal of Medicine* also supports research on and use of embryonic stem cells. In an editorial, it criticized the U.S. House of Representatives for voting to ban research on and use of the stem cells and noted that the ban "has the potential to put many critical future advances in medicine beyond the reach of patients in the United States."

The journal held that research with embryonic stem cells will continue, but that the research may be done outside of the United States rather than inside the United States. This puts the United States in the unaccustomed position of no longer being in the forefront of certain kinds of medical research, and it means that patients in the United States would be banned from receiving the most current medical care. The *New England Journal of Medicine* concludes, "No matter what Congress decides, such treatments will be developed somewhere in the world. Physicians and scientists in the United States should be at the center of the action, not on the sidelines."

The journal also said that it would, from now on, give papers about stem cell research a prominent place in its coverage and issued a call for new papers.

BioFact

Researchers at the University of Pennsylvania have found a way to turn mouse stem cells into egg cells. Previously, many scientists thought that it was impossible to do this. It is an important breakthrough because it means that ultimately there may be a way to do embryonic stem cell therapy without having to use existing eggs or embryos. One of the problems with stem cell cloning is finding eggs that can be used in the therapy. A stem cell could be turned into an egg, and then that egg could be turned into an embryo using normal cloning techniques.

Embryonic Stem Cells Are Superior to Adult Cells

Those who would ban therapeutic stem cell cloning claim that there's no need to use embryonic stem cells because adult stem cells can be used in their place. But scientists point out that there are many differences between embryonic stem cells and adult stem cells, and embryonic ones are far superior to adult ones for research and therapy.

As we've explained, embryonic stem cells have the ability to turn into any cell in the body, while adult stem cells can only turn into specific types. Additionally, embryonic stem cells are much faster-dividing than adult stem cells, which can be therapeutically beneficial, and are better suited to research than slower-dividing cells.

Adult stem cells are rarer and are difficult to identify and grow in the laboratory. Unlike embryonic stem cells, they do not replicate themselves indefinitely when grown in culture in the laboratory. They will not be nearly as usable as embryonic stem cells.

For all these reasons, embryonic stem cells hold out far greater promise than do adult stem cells for fighting disease.

The Celebrities Weigh In

Like it or not, we live in a celebrity-obsessed culture in which celebrities are looked to for advice on everything from political issues to sex, love, weight loss, and the meaning of life. Want to know who to vote for? A lot of people wish they could ask Madonna or Mel Gibson. Frightening thought, we know, but people actually listen to them.

So perhaps it is no surprise that celebrities have weighed in on the stem cell issue as well. But in this case, the celebrities who have paid attention to the issue actually have something useful to say, because they have been personally touched by it.

BioFact

Scientists have found what some call an "immortality gene" or "master gene" that appears to give embryonic stem cells their ability to continually multiply and also to be able to turn into any other cell in the body. They hope that eventually, through manipulating the gene, they may be able to turn any cell in the body into a gene with the same capabilities of embryonic stem cells. The gene was named "nanog" after the mythical land in Celtic lore Tir nan Og, whose inhabitants are believed to be always young and live forever.

The most heart-rending case is that of actor Christopher Reeve. Reeve was paralyzed in May 1995, in a horse-riding accident and is confined to a wheelchair. His spinal cord cells are damaged and his is the kind of injury that researchers believe may be able to be ultimately cured by stem cells—stem cells could be turned into spinal cord cells, they say, and those undamaged spinal cord cells could replace his damaged cells. With this combination, he might be able to walk again.

Reeve has spoken out in favor of embryonic stem cell research, and told *CNN*, "If you had the FDA (Food and Drug Administration) involved and everybody working together, I am positive in 10 years I'd be on my feet … I would not be sitting here in a wheelchair."

For those who say that it is somehow unethical to use embryonic stem cells, he says, "You really don't have an ethical problem because you're actually saving lives by using cells that are going to the garbage … I just don't see how that's immoral or unethical. I really don't."

Another celebrity who has weighed in on the issues is Mary Tyler Moore who has "juvenile" or Type 1 diabetes. There are those who say that embryonic stem cells could be turned into insulin-producing cells, and so help cure diabetes.

BioFact

Stanford University has taken a strong stand in favor of stem cell cloning—so much so that it has started a new institute devoted to it, called the Institute for Cancer/Stem Cell Biology and Medicine. The institute will focus on research into the basic biology of stem cells and on using stem cell research to treat diseases such as cancer, Parkinson's disease, diabetes, and neurodegenerative disorders.

In asking that the federal government fund embryonic stem cell research, Moore testified before Congress that "My 30 plus years of diabetes has led to visual impairment, painful neuropathy, the threat of limb loss from poorly healing foot wounds, and peripheral vascular disease which has started to limit how far I can walk."

She added, "One American dies from diabetes every 3 minutes. We owe [this to] children and the 16 million with the disease, to pursue all promising research avenues … including stem cell research."

In addition to lobbying Congress on the issue, she has co-founded a fund to support embryonic stem cell research.

The actor Michael J. Fox, who suffers from Parkinson's disease, also favors embryonic stem cell cloning and has also testified in Congress in favor of it. Parkinson's disease is one of the diseases that researcher believe may be able to be cured by the use of embryonic stem cells.

BioFact

Acting celebrities certainly aren't the only well-known people who favor embryonic stem cell research. Dr. Harold Varmus, former director of the National Institutes of Health, a Nobel Prize winner in Medicine, and currently President of the Memorial Sloan-Kettering Cancer Institute, testified before Congress in favor of funding embryonic stem cell research, and concluded, "It is not too unrealistic to say that this research has the potential to revolutionize the practice of medicine and improve the quality and length of life."

Why Therapeutic Stem Cell Cloning Should Be Banned

Those who oppose therapeutic stem cell cloning generally do not oppose it for any scientific reasons or a belief that the procedure is dangerous. In fact, even many opponents of the research agree that it may hold out medical benefits.

Opponents are against the procedure because they believe that a fertilized egg or an embryo is in fact a human being. Therefore, doing research on embryonic cells is a form of murder. In short, they oppose embryonic stem cell cloning for the same reasons that they oppose abortion.

The Catholic church, for example, has come out strongly in opposition to embryonic stem cell cloning. So have anti-abortion groups and activists, although some prominent anti-abortion politicians favor research into the technology. For example, prominent anti-abortion Senator Arlen Specter, a Republican from Pennsylvania, favors embryonic stem cell research, as does Republican Senator Orrin Hatch of Utah. Former First Lady Nancy Reagan, another abortion opponent, supports the research as well. Former President Ronald Reagan suffers from Alzheimer's disease, and there are those who believe that embryonic stem cells may hold a cure for it.

The opponents of therapeutic stem cell cloning wield a great deal of political clout. For example, they have managed to get bills passed in the House of Representatives outlawing new research, although as of this writing, the Senate has not yet passed a similar bill. In addition, they have managed to convince the American Heart Association not to fund research into embryonic stem cell research. The group will only fund research into adult stem cells and will not fund any research into embryonic stem cells.

Some opponents of the research and therapy also say that they worry that stem cell cloning is the first step toward cloning humans, which they oppose.

Federal Rules Involving Embryonic Stem Cell Research

The Federal government is a massive financial supporter of medical and scientific research, particularly at universities and non-profit research centers. Therefore, the government has enormous clout in deciding which technologies will be pursued and which will be ignored. As a practical matter, it is almost impossible for university researchers to pursue their research without federal funds.

President George Bush's decision on embryonic stem cell funding has had a major impact on the future of the technology and therapy. His decision, the logic of which can be very difficult to follow, has put a major crimp in embryonic stem cell research in the United States. And as we'll see later in this chapter, it's also led to a minor "brain drain" the form of some U.S. researchers heading for Europe.

On August 9, 2001, Bush decreed that the federal government will only fund research into embryonic stem cells for certain stem cell "lines," those in which the line was created before August 9, was made from an embryo that had been created for in vitro fertilization, that the donor was informed what the embryo was to be used for, and that no financial incentives were involved.

As a practical matter, this governmental stance set back embryonic stem cell research. It meant that scientists could only use certain stem cells in their research, and not any others. It also meant that there could conceivably be a shortage of embryonic stem cells available for research. Any new lines of embryonic stem cells could not be used for research. Scientists warned that as a result, medical research and therapy would be set back in the United States—and that the research would go on in other countries, and so the United States would no longer be in the forefront of this kind of technology.

BioSource

For a full explanation of President Bush's ruling on the use of federal funds for embryonic stem cell research, go to the Stem Cell Information site at http://stemcells.nih.gov/index.asp.

At the time of the announcement, the President and the National Institutes of Health (NIH) claimed that there would be more than enough stem *cell lines* to go around for researchers to use and so research would not be hampered. The NIH claimed that there were more than 70 lines of stem cells.

BioDefinition

A **cell line** is a self-propagating colony of cells from a single source. So an embryonic stem cell line would be a colony of embryonic stem cells that came from the same initial embryo and that continue to reproduce continually.

However, that has not been the case. In May of 2003, the Director of the National Institutes of Health, Elias Zerhouni, announced that in fact there were only 11 embryonic stem cell lines that could be used by researchers who wanted to be able to receive federal funds for their research.

Criticism of the President's Policy

In the nearly two years between Bush's announcement and Zerhouni's admission that only 11 stem cell lines were available, critics say stem cell research suffered greatly and things are only getting worse.

For example, George Dale of the Whitehead Institute at the Massachusetts Institute of Technology, spoke for many scientists when he testified before Congress saying, "The existing restrictions are keeping advances from being realized"—and he said that before Zerhouni revealed that only 11 stem cell lines were available.

Donald Kennedy, editor of the prestigious journal *Science*, wrote in an editorial, "It is clearly not sound policy to retain the current restriction," because it is hampering research.

The Senate panel that funds medical research held hearings on the matter in the spring of 2003, and senators and scientists were unanimous in their disapproval of the policy. Senator Specter, a head of the panel, said, "The hands of the scientists shouldn't be tied in any way." And Senator Tom Harkin said, "Every time I see someone with Parkinson's, every time I see someone with a spinal cord injury, or I see someone with Alzheimer's, I ask the question: 'Why aren't we moving more aggressively on this?'

Specter asked the NIH's Zerhouni to identify a nongovernmental scientist who would testify in favor of President Bush's policy. But the NIH could not identify a single scientist not on the federal government's payroll who supported the policy. So none testified.

The United States Falls Behind

There are signs that the predictions of those who say that the United States will fall behind in embryonic stem cell research because of Bush's decision, may have started to come true. Swedish researchers and researchers in Singapore have been able to create new embryonic stem cell lines that federally funded U.S. researchers are banned from using because those cell lines were created after August 2001.

The stem cell lines in the United States use mouse cells as a way to spur their growth, but the researchers elsewhere have been able to create and grow stem cells free of contamination from mouse cells. Some people worry that the mouse cells might contain pathogens such a mouse retroviruses. If that were the case, one wouldn't want to implant the virus-contaminated cells in people because of the potential medical dangers. However, because of Bush's restrictions, U.S. researchers will not be allowed to use the uncontaminated cells in their research if they want to receive federal funds.

BioFact

President Bush's ruling did not outlaw research into embryonic stem cells created after August, 2001—it can still be legally done. But as a practical matter, little research is being done with those stem cells because so much research is federally funded. However, there is a movement in the U.S. Congress to outlaw all embryonic stem cell research, even by private biotech companies. A bill to do that has passed the House of Representatives, but not the Senate.

Some American scientists have even started to take positions overseas because of the U.S. restrictions. For example, the magazine *The Scientist* reported that a young American biologist named Diana Devore moved to Australia to become chief operations officer of the Australian government-funded National Center for Stem Cell Research because U.S. restrictions made research in the United States difficult.

The States Step In

Where the federal government fears to tread, states step in. States that have strong ties to the medical and bioresearch community have been trying to pick up the slack caused by Bush's policy. The state of California, for example, has passed a bill supporting embryonic stem cell research, and a variety of California institutions are funding the research using nonfederal funds. For example, Andy Grove, president of the chip company Intel, gave a $5 million grant to the University of California San Francisco, to establish a Developmental and Stem Cell Biology Program. Stanford has similarly set up an institute.

Massachusetts and several other states are looking to do the same. At hearings at the Massachusetts state house, a variety of researchers testified in favor of a bill similar to California's. At the hearing's most moving moment, Travis Roy spoke. In his first game as a Boston University freshman hockey player, Roy was paralyzed. He testified: "Whenever you hear people talk about curing paralysis, you always hear the same words: To walk again. But it's so much more than that. It's to feel again, to have control of bowel and bladder again. It's to have sensation and to have normal sexual functioning. Stem cells are my biggest hope."

Similar bills have been introduced in New York, New Jersey, Tennessee, Rhode Island, and Maryland.

The Least You Need to Know

- Those in favor of therapeutic stem cell cloning say that it can cure many diseases, such as Parkinson's disease and diabetes, and cure many other conditions, including spinal cord injuries.
- Anti-abortion activists are against therapeutic stem cell cloning because they say that it destroys embryos, and they consider embryos to be human life.
- Many opponents of abortion, such as prominent senators and former First Lady Nancy Reagan, are in favor of therapeutic stem cell cloning because it can save lives.
- President Bush has banned the use of federal funds for therapeutic stem cell cloning except in instances in which the stem cell "lines" were created before 2001.

- Many scientist claim that Bush's ban has set back medical research in the United States, and say that other countries will take the lead in the research.
- States such as California, Massachusetts and others have already passed or are considering passing laws that would help spur therapeutic stem cell cloning in those states.

Chapter 15

The Feds Speak Up

In This Chapter

- Which federal agencies regulate cloning and biotechnology
- How the U.S. Department of Agriculture regulates transgenic crops
- The role that the Food and Drug Administration plays in regulating biotechnology
- What the critics say about government regulation of biotechnology
- What the backers of current regulations say about the role government plays

We have laws and regulations that cover just about every aspect of our lives. Where you can park and where you can't park. At what temperature supermarkets must store ice cream. At what height you can construct new buildings in certain neighborhoods. Where you're allowed to cross streets. At what time 13 year-olds must go to sleep … oh, forget that one. That's a regulation in the Cambridge Gralla household and it's breeched far more often than it's obeyed.

Not surprisingly, a variety of federal laws and regulations cover cloning and biotechnology. We don't have room to cover them all here—to do that would take several very thick books, not a single chapter, and it would be very boring. But in this chapter we'll cover the most important rules, regulations, and recommendations—and we'll discuss some of the disagreements and controversies around whether government is doing an adequate job of regulating the technology.

Who Regulates Cloning and Biotechnology?

The United States is a crazy quilt of government bodies, starting with local governments at the town and city level, on up to county government, state government, and federal government—and often, there are government bodies in between, such as regional transportation and planning authorities.

But when it comes to regulating cloning and biotechnology, federal government is the place that counts. That's where the most important regulations can be found, and where enforcement takes place.

Given the amount of red tape involved in almost anything to do with the feds, and the complexity of biotechnology, it's no surprise that no single agency regulates bioengineering. In fact, there is a complex patchwork of regulations and agencies involved.

BioFact

Although the federal government is the primary government regulator of biotechnology, a number of states get into the act as well. For example, several states have regulations concerning the testing or use of bioengineered plants and animals. North Carolina has passed a Genetically Engineered Organism Act that requires a permit for field testing of a genetically engineered plant, animal, or microbe. The act's regulations are designed to work with federal agencies that also regulate genetic engineering, so that researchers need not jump through too many hoops in order to do their work.

The Agencies at a Glance

Here are the primary agencies that regulate biotechnology:

- **The Food and Drug Administration (FDA)** The Food and Drug Administration regulates food additives and new foods, as well as drugs. So it regulates bioengineered foods introduced into the food supply, as well as bioengineered drugs.
- **The U.S. Department of Agriculture (USDA)** The main branch of the USDA that regulates biotechnology is the Animal and Plant Health Inspection Service (APHIS). APHIS is charged with safeguarding agriculture from diseases

and pests. The agency regulates the field-testing, importation, and interstate transport of genetically engineered microorganisms and plants. It also regulates animal vaccines, and therefore regulates any bioengineered vaccines, including those delivered by bioengineered means. A bioengineered vaccine might be produced by transgenic plants and then eaten by animals.

BioFact

The FDA regulates food labeling and there has been a call by consumer groups to have any genetically modified (GM) foods carry a label that identifies the food as having been genetically modified. However, the FDA has refused to do that, claiming that there is "no basis for concluding that bioengineered foods differ from other foods in any meaningful or uniform way, or that, as a class, foods developed by the new techniques present any different or greater safety concern than foods developed by traditional plant breeding." We'll take a closer look at this issue later in the chapter.

- **The Environmental Protection Agency (EPA)** This agency is responsible for the protection of the environment and human health as affected by the environment. Of most relevance to biotechnology is its regulation of pesticides. Therefore, any plants that have been genetically modified to produce their own pesticides are regulated by the EPA.
- **The National Institutes of Health (NIH)** This agency, which administers billions of dollars in federal research dollars, regulates recombinant DNA research for any institution that receives NIH money for that research. Private industry, however, does not have to comply with NIH guidelines if they do not receive NIH money for their research.

BioSource

If you're a government junkie and want to see what the various government agencies have done, or plan to do about bioengineering, you can visit the portions of their websites devoted to bioengineering. If you'd like to see all the actions and proposed actions that the FDA has taken with regard to biotechnology, go to www.cfsan.fda.gov/~lrd/biotechm.html#label.

For a biotechnology overview for the U.S. Department of Agriculture, go to www.aphis.usda.gov/biotech/usda_biotech.html and to www.aphis.usda.gov/brs/.

For the EPA, go to www.epa.gov/pesticides/biopesticides/index.htm for information on how the agency regulated biopesticides, and for genetically engineered microbes, go to www.epa.gov/opptintr/biotech/index.html.

And for the National Institutes of Health, head to www4.od.nih.gov/oba/ for the agency's Office of Biotechnology Activities.

The federal government influences bioengineering and cloning in ways other than regulation, however. The government spends billions of dollars a year for research, and the way it decides to spend that money helps determine the direction of research. For example, the federal government limits in certain ways money spent on therapeutic stem cell cloning, and some critics say that because of that, some research into the technology has moved overseas.

What Laws Regulate Cloning?

Amazingly enough, as of this writing there are no rules that specifically regulate the cloning of human beings, or that ban the cloning of human beings. As a practical matter, human cloning is still well beyond the capabilities of science, but in theory at least, there are no federal laws banning the practice. As long as the other biotechnology regulations are followed, all systems are go.

The reason for that, not surprisingly, has to do with politics. There are few things that Democrats and Republicans agree upon, but they generally all agree that human cloning should be banned. As of this writing, though, they haven't been able to pass a law banning the practice. Whenever the issue comes up, Republicans tack on an amendment that would also outlaw therapeutic stem cell cloning, which does not involve cloning human beings. Instead, it involves the use of stem cells from embryos for medical research therapy—many people have said that stem cells may be a kind of "miracle cure" for many diseases. However, many Republicans oppose stem cell use because it involves cells from embryos, although from embryos that were used in fertility clinics and would otherwise have been discarded. For more information about therapeutic stem cell cloning, turn to Chapters 8 and 14.

> **BioFact**
>
> Some countries have already outlawed human reproductive cloning. The United Kingdom, for example, bans human reproductive cloning, having passed a law against it late in 2001. The UN, however, has been unable to pass a treaty banning cloning, largely because the United States and the Vatican have pushed for a worldwide ban on therapeutic stem cell cloning as well as human reproductive cloning.

Democrats generally favor therapeutic stem cell cloning, and because the anti-stem cell cloning amendment keeps getting tacked onto the law to ban human cloning, no law has yet been passed. There's a possibility that by the time you read this that may have changed.

The FDA and Biotechnology

The FDA is the main federal agency responsible for regulating new foods, food additives, and drugs, so it's right in the middle of regulating transgenic, bioengineered foods, and drugs. Given the complexity of the federal government's bureaucracy, however, there are some twists to that regulation. When it comes to meat and poultry, the USDA has primary authority. But when it comes to the genetically engineered animal growth hormones bST and pST, it's the FDA that rules because the FDA regulates animal drugs. Got it? If you do, you're one of the few people in the world who does.

In 1992, the FDA issued its primary guidelines regarding genetically modified (GM) foods. It ruled that generally any new plant created by genetic engineering would be treated no differently than a new plant created by conventional means, such as by the normal plant breeding that has gone on for millennia.

However, it also ruled that it would require a review of any new GM food if there were any safety issues that the new food might pose, such as the following:

- If it contains proteins, such as a peanut protein, that might cause allergic reactions in people.
- If it may cause unexpected genetic effects.
- If it significantly alters the level of important nutrients.
- If it contains significantly higher levels of known toxicants than are found in non-GM varieties of the same species.
- If it contains new substances that are significantly different than those currently found in foods.
- If it contains genes that may reduce the effectiveness of antibiotics.
- If it is a plant designed to make non-food products, such as pharmaceuticals or other chemicals.
- If it is GM animal feed in which there are significant changes in nutrients or toxicants.

CAUTION

BioWarning

One potential danger with biotechnology research is that plants or animals from the laboratories somehow enter the food supply without proper trials. In one instance, bioengineered pigs at the University of Illinois were sold as food and apparently eaten by unwary consumers, says the FDA. However, the agency holds that there were no dangers from eating the animals. The pigs were the offspring of transgenic pigs that had been engineered so that proteins would be produced in their milk. Researchers say that the offspring did not contain the transgene of their parents and that the slaughtered pigs were so young that they were not yet lactating. Even so, because the pigs were released into the food supply improperly, the FDA said that it would ultimately take action.

As a practical matter, companies frequently check with the FDA before releasing a new GM food because of the serious economic and legal consequences they would face if they were forced to withdraw a food from the market after they had introduced it. In addition, the FDA has proposed, but not put into effect, that all GM foods and animal feeds undergo a notification process in which the companies will notify the FDA before the product is released. However, it would not require testing or screening of the product.

The FDA is also responsible for food labeling, and as you'll see later in this chapter, it has been involved in a controversy because some consumer and environmental groups want it to require that all GM foods be labeled as such.

The EPA and Biotechnology

The EPA regulates pesticides and herbicides, and so it also regulates any pesticides produced in transgenic plants as a result of genetic engineering—for example, Bt corn that has been genetically engineered to create a natural pesticide so that no chemical pesticides need be used. It also regulates any herbicide-resistant plants created as a result of genetic engineering. Farmers would use herbicide-resistant plants so that they can spray fields with herbicides and kill only weeds, and not disturb the transgenic crop.

BioFact

As of this writing, all of the transgenic plants registered with the EPA that produce their own pesticides do so using Bt. The Bacillus thuringiensis (Bt) bacterium produces a crystal-like Bt protein that kills certain insects but is harmless to people and animals. Researchers isolated the gene that produces Bt and have implanted it in a variety of crops, notably corn. (For more details, turn to Chapter 23.) The first Bt bioengineered plant registered with the EPA was a potato plant in 1995. Between 1995 and June 2003, a total of nine transgenic plants that produce their own pesticides using Bt were registered. Six of them were corn, two were potatoes, and one was cotton.

In one of those surreal, "it-can-only-happen-in-Washington" rules, the EPA actually regulates only the protein and the genetic material that produces the protein in a transgenic plant, not the plant itself. That falls to the USDA.

That's not the only overlap between the EPA and the USDA. The USDA regulates field trials of transgenic plants. If those plants have been genetically modified to produce their own pesticide for example, they will be regulated by both the EPA and the USDA.

There are those who say that this kind of overlap means there's not an effective way to regulate biotechnology because there is no single agency that handles all aspects of the technology. Others argue that it makes sense to have agencies with the proper expertise regulate different aspects of bioengineering. We'll look at the issue in more detail later in this chapter.

The USDA and Biotechnology

The Animal and Plant Health Inspection Service (APHIS) is the primary division of the USDA that regulates agricultural biotechnology. It's charged with safeguarding U.S. agriculture from diseases and pests. In addition, it regulates the field testing of GM foods and organisms, as well as their importation and transportation between states.

APHIS reviews the proposed field tests to make sure that the proposal is safe, and then after the testing is done, determines whether the genetically modified product can be grown commercially. It reviews about 1,000 field tests a year for GM crops.

APHIS is supposed to make sure that in the field tests, the GM crops cannot escape from the area where they're being tested, and that pollen from the plants can't escape as well. For example, APHIS regularly requires that the plants are transported to the test site in special enclosed containers. Depending on the plant being tested, it may be required that flowers from the plants are bagged in order to stop cross-pollination with nearby non-GM crops. The spring after the crop is harvested, the site is typically treated with herbicides, and then the site is monitored to make sure that no GM crops remain.

BioSource

Using the federal Freedom of Information Act, anyone who wants can get copies of the APHIS permit applications that must be filed when genetically engineered plants are field-tested—and copies of the environmental assessments are available as well. To request them, contact the Freedom of Information Act Coordinator; USDA, APHIS, LPA, PI; 4700 River Road; Riverdale, MD 20737.

Are Agencies Regulating Biotechnology Properly?

Some consumer and environmental groups contend that the federal agency isn't properly regulating biotechnology, particular GM crops. They say that too many different agencies are involved, and it is too easy for things to fall between the cracks. Some contend that federal oversight of field tests has been lax and that this may ultimately lead to serious health and environmental consequences.

BioFact

Different countries have different regulations regarding biotechnology and cloning and they don't always agree with one another. Because of the emotion involved with the cloning issue, it's unlikely that there will be an international agreement regarding cloning any time soon. However, there has been an attempt to come to an international agreement over how bioengineered foods and animals can be introduced from one country to another—it is called the Cartagena Protocol on Biosafety. As of this writing, a little over 50 countries have signed the protocol; however, many countries that make use of transgenic crops have not signed it, including the United States, Argentina, China, and a number of other large countries. For information about the protocol, go to www.biodiv.org/biosafety/ratification.asp.

For example, transgenic corn that was bioengineered in a field test to produce a pharmaceutical ended up in soybeans that were going to be sold for human and animal consumption. Before the crop was sold, the mix-up was discovered, and the crop was destroyed.

The Case of ProdiGene, Inc.

Here's what happened. ProdiGene, Inc., a Texas-based biotech firm, used a test plot to grown transgenic corn that had been genetically altered to produce a pharmaceutical protein. The following season, the plot was used to grow soybeans that were going to be sold for human and animal consumption. Seeds from the transgenic corn had been left behind in the field, however, and so the transgenic corn that produced the pharmaceutical grew along with the soybeans. The corn was harvested along with the soybeans, and together they were stored in a grain elevator with other soybeans—all told, the grain elevator contained nearly 500,000 bushels of soybeans and a small amount of the transgenic corn.

The mistake was discovered, and all 500,000 bushels of soybeans—along with the small amount of transgenic corn—were destroyed. APHIS fined ProdiGene $250,000 and forced the company to also pay for the cost of the cleanup.

BioFact

The government influences biotechnology not just through regulations, but through the research it pays for. One of the most significant actions it has taken regarding biotechnology was the controversial decision made by President George W. Bush to support therapeutic cell cloning research in only a very limited way. Federal funds can only be spent on a limited number of stem cell "lines" created before August, 2001. For more information, see Chapter 14.

APHIS had already been considering stricter regulations of these kinds of crops. These include requiring that they be grown further from other food crops, and that no feed or feed crops be grown on the testing site the following season. The discovery of the problem spurred the agency to issue the new regulations.

In one of the rare instances in which private industry has actually asked that a federal agency regulate it more strictly, Fred Yoder, a corn farmer and president of the National Corn Growers Association said, "I never thought I'd be one to ask for more requirements, but we did. We don't want to see an AIDS vaccine show up in a box of cornflakes. And with these regulations, that will never happen. If we don't do it right, we won't have this technology."

A Report on APHIS

The National Academies of Sciences Natural Research Council did a comprehensive study of the way that APHIS regulated biotechnology, and in a February 2002 report concluded that the agency "should more rigorously review the potential environmental effects of new transgenic plants before approving them for commercial use." The report also concluded that "the public should be more involved in the review process and that ecological testing and monitoring should continue after transgenic plants have entered the marketplace."

Some environmental and consumer groups say that the report did not go far enough, and even say that the USDA, through APHIS, should not be the agency to regulate transgenic crops. They claim that there is a conflict of interest in having the USDA regulate the crops because the agency's primary mission is to promote agriculture. Therefore, it would naturally want to promote new transgenic crops, not limit their use. Jane Rissler of the Union of Concerned Scientists in Washington, D.C. told the Pew Initiative on Biotechnology, "APHIS is operating under a weak statute and using that statute weakly. The agency isn't doing the kind of regulation that it should be. There's a bias in favor of technology. And I don't think the USDA can be a good regulator of products that it itself promotes vigorously."

On the other side, the biotechnology industry claims that the agency is doing a good job—after all, nobody has ever been harmed by a GM food and nearly everyone in the United States has eaten one. "This report concludes that the existing regulatory system is founded on sound science," Val Giddings of the Biotechnology Industry Organization told the Pew Initiative on Biotechnology. "We have a functioning and efficient regulatory system in the United States. As good as it is, nothing is perfect, and USDA is to be commended for asking the Academy for this report."

A year after the report, APHIS strengthened its guidelines for plants that have been genetically engineered to produce pharmaceutical and industrial compounds. For example, it requires that no corn can be grown within a mile of GM corn created that produces pharmaceutical and industrial compounds, it increased the number of site inspections, and put in other restrictions.

BioSource

For a copy of the National Academies of Sciences report on federal regulation of transgenic plants, go to www.nap.edu/catalog/10258.html?onpi_newsdoc022102. You'll be able to read the report online, and print out the report, but only a page at a time. You can also order a copy of the full report from the website. You can buy an electronic version for $30, or a hardcover copy for $39.96.

The GM Labeling Controversy

Perhaps the most contentious issue having to do with bioengineering and government regulation is whether GM foods should be labeled—in other words, whether genetically modified foods should have labels identifying them as such. The FDA and the food industry say that there is no need for them; some consumer and environmental groups say that there is a need.

The FDA doesn't require labels on GM foods because it claims that GM foods are equivalent to non-GM foods and are safe. If, however, at some point a GM food was substantially different than a non-GM food—if it was significantly nutritionally different or contained potential allergens—it would require a label for that specific food. As of yet, no foods require that label.

One of the many consumer groups that favors labels is the Consumer's Choice Council, based not on a belief that GM foods are dangerous, but that consumers should be able to know what they are buying.

"We don't take a pro or con view of biotechnology," said Cameron Griffith, Director of Government and Labor Relations for the Consumer's Choice Council. "But we do think there is a right to know and if you believe biotechnology is a good thing, then you should be willing to label the food. Consumers have the right to know."

Among the many industry and food groups that oppose labeling is the Grocery Manufacturers of America. The group contends that labels would impart no useful information to consumers. Because GM foods are as safe as non-GM foods, the group says, it would be misleading to label GM foods because it would draw attention to GM foods and in some way imply that the food is not as good or is more dangerous than non-GM food.

The FDA has worked on a voluntary labeling scheme in which food producers can label foods as GM-free, so that they can avoid GM products by buying food with that label. Draft guidelines have been looked at, but the scheme is not yet in place. Some food products already include a claim of being GM-free, but that claim is not FDA-endorsed.

BioFact

In November 2002, Oregon citizens voted on a measure, Ballot Measure 27, which would require that all GM food sold in the state carry a label identifying it as GM food. The voters spoke loudly and clearly: 73 percent of them voted against the measure and so it did not become law.

The Least You Need to Know

- Biotechnology is primarily regulated by three government agencies: the Food and Drug Administration (FDA), the Environmental Protection Agency (EPA), and the U.S. Department of Agriculture (USDA).
- As of this writing, the United States has not banned human cloning because of political disagreements between the Republicans and the Democrats about therapeutic stem cell cloning.
- The FDA regulates bioengineered foods introduced into the food supply, as well as bioengineered drugs.
- The USDA regulates the field-testing, importation, and interstate transport of genetically engineered microorganisms and plants.
- The EPA regulates bioengineered pesticides that are created by transgenic plants to produce their own pesticides.
- Some consumer and environmental groups favor putting labels on genetically modified foods, but government and industry groups say that to do so would mislead consumers by implying those foods are in some way dangerous.

Chapter 16

Religion and Cloning

In This Chapter

- What Genesis might say about cloning
- The Catholic Church's stand on cloning
- Protestant denominations and cloning
- Judaism and cloning
- What Islam, Hinduism, and Buddhism say about cloning

Cloning presents one of the great ethical debates of our era. How we and our laws handle cloning and related biotech issues will, to a great extent, determine the kind of society we and our children live in.

Religion has played a central role in the debate so far. Churches are in the forefront of the discussions, and many people believe that religious groups have to a great extent determined President George W. Bush's decision to ban funding for embryonic stem cell research, with certain stem cell lines exempt.

Religions vary dramatically in their views about human cloning and embryonic stem cell cloning. Some groups want outright bans on both technologies, some want a cloning ban, but want to allow embryonic stem cell cloning, and some want to allow both. There is not even necessarily agreement within each religious group.

In this chapter, we'll see what religions and religious leaders have to say about human cloning and embryonic stem cell cloning.

Religion's Unique Perspective on Cloning

Religion is often portrayed as the enemy of science, and in the past that has certainly at times been the case. In the seventeenth century, for example, Galileo was found guilty of heresy by the Catholic Church for his work that declared that the sun, not the earth, was the center of the solar system and that the earth revolved around the sun, rather than the sun revolving around the earth. He was imprisoned during the last years of his life as a result of being found guilty. And in much more recent times, in the famous Scopes "monkey trial," a schoolteacher was found guilty in 1925 in Tennessee for teaching evolution.

BioFact

It took more than 350 years after Galileo's death for the Catholic Church to admit that it made an error. On October 31, 1992, Pope John Paul II admitted that errors had been made by the theological advisors when Galileo was convicted of heresy. It took a study committee appointed by the Pope 13 years to come up with the finding.

But religion is not necessarily the enemy of science, at least not in modern times. And when it comes to cloning, the thinking of many religious leaders tends to be in line with the general populace, as well as with many scientists. Many religious groups express mainstream thinking, not anti-science thinking, when it comes to cloning.

BioSource

If you'd like to find out more about the odd Raelian UFO cult, go to www.raelian.org. Be prepared for one of the oddest visits to cyberspace—you'll read not only about UFO cloning, but also about crop circles, and even how to get yourself excommunicated from the Catholic Church.

In fact, many of the reasons that religious groups oppose human cloning are based as much on societal and medical concerns as they are on purely theological ones. The reasons many support cloning are based on societal and medical concerns as well.

In this chapter, we're not going to get into a debate about how you define a religious group. The Raelians, for example, is a movement founded by a former French racing car journalist that believes that humans were bioengineered by extraterrestrial beings, and not only supports cloning, but claims to have already cloned a baby. It calls itself a religion and supports cloning.

In this chapter, we'll cover only mainstream religious groups and their views about cloning and will not cover groups like the Raelians.

Genesis and Cloning

For Judaism and Christianity, much of the religious debate around cloning starts at the very beginning of the Judeo-Christian Bible—in the first book of the Bible, Genesis, which describes the creation of the universe and of mankind. There are several passages that are the most important, beginning with chapter 1, verse 1:

> In the beginning God created the heaven and the earth.

The next most important passages relating to cloning are chapter 1, verses 27–30:

> And God said, Let us make man in our image, after our likeness: and let them have dominion over the fish of the sea, and over the fowl of the air, and over the cattle, and over all the earth, and over every creeping thing that creepeth upon the earth.
>
> So God created man in his [own] image, in the image of God created he him; male and female created he them.
>
> And God blessed them, and God said unto them, Be fruitful, and multiply, and replenish the earth, and subdue it: and have dominion over the fish of the sea, and over the fowl of the air, and over every living thing that moveth upon the earth.
>
> And God said, Behold, I have given you every herb bearing seed, which [is] upon the face of all the earth, and every tree, in the which [is] the fruit of a tree yielding seed; to you it shall be for meat.
>
> And to every beast of the earth, and to every fowl of the air, and to every thing that creepeth upon the earth, wherein [there is] life, [I have given] every green herb for meat: and it was so."

BioSource

The excerpt from Genesis quoted here is taken from the King James version of the Bible. Other editions may vary in their wording somewhat, but the meaning is the same.

There are a number of themes here that help form Judaism and Christianity's reaction to cloning. Start at the very beginning. There are some who say that the first line of the Bible implies that only God can create life. In fact, this idea goes beyond religion. One of the most common reasons that laypeople oppose cloning is that they believe we shouldn't "play God."

Some also point to the passage in which it is said that man is created in God's image. Some religious leaders use that as a reason to ban cloning—when a human being is cloned, some contend, it means the human being is being made in man's image, not in God's image. Even those who don't take the Bible literally point to this passage and use it as metaphor. By saying that man is made in God's image, it means that every individual is unique with his own purpose in life. Cloning, they say, takes away that individuality.

The passage where mankind is told to, "Be fruitful, and multiply, and replenish the earth, and subdue it," poses something of a conundrum for some religious thinkers because it seems to imply that cloning, because it concerns procreation, should be condoned by the Bible. Because it tells mankind to "subdue" the earth, it also seems to imply that man in some way stands outside of nature, and can master nature.

BioSource

If you're looking for a copy of the entire King James Bible, you can find it online free. Go to etext.lib.virginia.edu/kjv.browse.html, part of a collection of free electronic books made available by the Electronic Text Center at the University of Virginia. You can search through the entire Bible or sections of the Bible for words or phrases at the site. A search of the Bible for the word "clone" shows that the term was never mentioned.

Generally, though this passage is interpreted to mean that man is given a kind of "responsible dominion" over the earth—in other words, he is entrusted with a kind of stewardship over the earth. Cloning, they say, falls outside the bounds of that kind of stewardship.

The Catholic Church and Cloning

The Catholic Church has probably been the most visible religious group involved in the cloning debate, possibly because it has taken very public and very unequivocal stands against not only human cloning, but also against therapeutic stem cell research involving embryos. Additionally, the church is governed in a "top-down" manner, with a single central authority deciding on religious matters. In other religions, such as Judaism, there is a multiplicity of allowed viewpoints. However, the Catholic Church speaks with a single, official voice and that single voice is more likely to be heard than a group of voices that might not agree with one another.

Part of the church's opposition to cloning stems from its general beliefs about reproduction and sexuality. The church opposes all methods of "artificial" birth control, such as condoms, pills, diaphragms, sterilization, any other method other than refraining from sexual acts, or using the "rhythm" method (not engaging in sex during the times of the month when a woman is most likely to be fertile). It also opposes in vitro fertilization, or any similar methods for helping women become pregnant.

BioSource

The church document, called an encyclical, that detailed once again the Catholic Church's opposition to birth control was issued in 1968 and titled "Humanae Vitae" (Latin for "Human Life"). The church contends that contraception is a violation of "natural law," which it says is the design God had in mind for the human race. The purpose of sex in natural law, according to the Catholic Church, is solely for procreation. Any pleasure that people might feel from it is only an additional benefit. Sexual pleasure becomes unnatural and harmful when it is separated from the basic purpose of sex, which is procreation.

Considering that the Catholic Church opposes virtually so many technologies relating to contraception and conception, including abortion, it is not a great surprise that it also opposes human cloning. But there are a number of issues unique to cloning that draw the church's opposition.

The church's stand on cloning is unambiguous. It says that "every possible act of cloning humans is intrinsically evil." In 1987, a document known as "Donum Vitae" was issued, which said that attempts to create "a human being without any connection with sexuality through 'twin fission', cloning or parthenogenesis are to be considered contrary to the moral law, since they are in opposition to the dignity both of human procreation and of the conjugal union." This very much mirrors the church's reason for opposing birth control as well.

There are additional reasons that the Catholic Church opposes cloning, however, and they are not necessarily based solely on religious teachings. In a statement of the Catholic Leadership Conference on Human Cloning in 2001, the church said that since a cloned child would carry only the DNA of one parent, the child's rights would be violated in a variety of ways. They noted that since no current law covers who would be responsible for the child, the child "would be left in a precarious position." It also noted that clones are often born with serious medical problems, another way in which the child's rights would be violated.

BioFact

Although the hierarchy of the Catholic Church opposes therapeutic stem cell cloning, that does not mean that all Catholics agree with the church. For example, many prominent Catholic politicians, such as Senators Edward Kennedy and John Kerry of Massachusetts favor federal funding for therapeutic stem cell research.

The church opposes therapeutic stem cell cloning as well as human cloning, and makes no real differentiation between the two. For example, the statement of the Catholic Leadership Conference on Human Cloning in 2001, "rejects as invalid the distinction drawn between therapeutic cloning and reproductive cloning". The Catholic Church teaches that an embryo is a human life (which is why it opposes abortion), and opposes stem cell cloning because embryos would be destroyed during the process of stem cell cloning. Although the church recognizes that stem cell research may help save human lives, it claims that killing embryos is evil and the ends do not justify the means.

Protestantism and Cloning

The stand of various Protestant churches tends to be more nuanced than the stand of the Catholic Church on the issue. Because there are so many different branches of Protestantism, there are many different views about cloning. Unlike with the Catholic Church, there is no single central authority. There is no possible way to cover all the views, so in this section, we'll cover the mainstream view of most churches.

As a general rule, Protestant churches oppose human cloning, but there is a difference of opinion about therapeutic stem cell cloning.

Evangelical Denominations

Commonly, evangelical denominations oppose both human cloning and therapeutic stem cell cloning, and conservative Protestant denominations tend to oppose both as well. The evangelical and socially conservative Southern Baptist Convention, for example, which has 16 million members in more than 40,000 churches nationwide, issued a statement in 2001 asking that both procedures be banned.

The resolution says that cloning humans "does not meet biblical standards for procreation in which children are begotten, not made." It claims that the Bible "declares that children are a gift from the Lord (Psalm 127:3–5) and are to be the offspring of a husband and wife (Genesis 1:27–28; 2:24; 9:1–2), not the result of asexual replication."

> **BioSource**
>
> For the full resolution of the Southern Baptist Convention on human cloning and therapeutic stem cell cloning, go to www.sbcannualmeeting.org/sbc01/sbcresolution.asp?ID=2.

Beyond that, it opposes both therapeutic stem cell cloning and human cloning for reasons that very much mirror those set out by the Catholic Church, saying that as soon as an egg fertilizes a sperm, a human being is created.

"A human embryo is a very young human being and nothing less," Richard Land, the president of the Southern Baptist Ethics and Religious Liberty Commission said in a statement in 2001. Since then, he has expanded his attack on all forms of cloning, claiming that human cloning could lead to the establishment of clone plantations creating slaves that could be sold to the highest bidder.

BioFact

Highly committed white evangelical Protestant Christians are opposed to therapeutic stem cell cloning far more than other groups, including mainstream Protestant groups, Catholics, African Americans, and Hispanics. A nationwide survey in 2002 by the Pew Research Center and the Pew Forum on Religion and Public Life found that the most religious white, evangelical, Protestant Christians oppose federally funded therapeutic stem cell cloning by a margin of 58 to 19 percent. By way of contrast, mainstream white Protestants favored it by a margin of 59 to 27 percent, African Americans favor it by 48 to 37 percent, and Hispanics favor it by 49 to 39 percent.

Other Protestant Denominations

The views of other Protestant denominations on cloning tend to mirror whether those denominations tend to be socially conservative or socially liberal. Many of the liberal or centrist denominations oppose human cloning, and favor therapeutic stem cell cloning, or take no stand on therapeutic stem cell cloning.

For example, the liberal United Church of Christ opposes human cloning and supports therapeutic stem cell cloning. The church does not consider that embryos are human beings, and so unlike the Catholic Church and socially conservative churches, does not favor a ban on cloning because it may kill embryos.

When it comes to the reasons for opposing human cloning, the church's response has more to do with issues of social justice than it does purely with scripture. Ronald Cole-Turner, who has been a spokesperson for the church on issues relating to cloning, notes in the book *Human Cloning: Religious Responses* that the church's opposition to cloning stems from opposition to devoting so many resources to the privileged and wealthy, while the poor of the world starve.

He writes:

> When the world groans with hunger, when children are stunted from chronic malnutrition, when people die of famine by the thousands every day … the development of any more technologies to suit the desires of those who are relatively privileged, secure, and comfortable seems to fly in the face of fundamental claims of justice.

As for therapeutic stem cell cloning, Cole-Turner has testified before the National Bioethics Advisory Commission, saying that the church believes that therapeutic stem cell cloning should be allowed and that the government should support that research with federal dollars.

BioSource

For the full statement of the United Church of Christ's support for stem cell cloning, go to www.ucc.org/synod/resolutions/res30.htm. For Ronald Cole-Turner's testimony before the National Bioethics Advisory, go to www.ucc.org/justice/stemcell/paper.htm.

The United Church of Christ has also taken an official stand on therapeutic stem cell cloning in its General Synod. The statement noted the enormous potential benefits of therapeutic stem cell research to help cure diseases such as Parkinson's, Alzheimer's, juvenile diabetes, and heart disease. It then went on to say that such research is very much in keeping with the Christian tradition because, "Jesus set an example, by his ministry of healing and caring for the sick and disabled, challenging us to follow his example by supporting the healing and caring ministry in our own day."

Judaism and Cloning

Judaism is made up of many different synagogues and movements, and there is no central authority that rules on religious and social matters such as cloning. Because of that, it is very difficult to summarize Judaism's stance on cloning and therapeutic stem cell cloning. Beyond that, many synagogues and movements simply have not taken a stand on the issue.

There is a great divergence of opinion in the Jewish community over cloning and therapeutic stem cell cloning. As we'll see, some rabbis and religious figures have even said that Jewish law allows human cloning.

The Jewish tradition generally places a strong emphasis on social justice and alleviating human suffering. Because of the travails the Jewish community has undergone through the centuries, it is particularly concerned with the suffering of others. As you'll see, this means that quite a few Jewish thinkers support therapeutic stem cell cloning, and this even leads to some of their support for human cloning as well.

Judaic View of Human Cloning

There is no unanimity among Jewish thinkers as to whether human cloning should be allowed. Many oppose it, and yet there are a surprising number who would allow it as well.

Those who oppose cloning do so for reasons already outlined in the opposition of Christian churches—that it would allow man to "play God," and that the technique itself could cause damage to child who will be born as a result of cloning. They also point out the various societal and familial issues.

However, some Jewish thinkers have come out in favor of human cloning, although with certain caveats. They say that it should only be done if it could be perfected so that there would be no dangers to mother or child. If that condition were met, some say that cloning should be allowed.

BioFact

Was Eve the first human clone? That provocative thought was raised by Rosalie Ber, an international lecturer on bioethics and head of the Medical Education Department at Haifa's Technion Israeli Institute of Technology. While not saying that Eve was actually cloned, she has pointed out that the Bible claims that Eve was taken from Adam's rib. So that those who take the Bible literally rather than metaphorically could actually believe, in some way, that Eve was cloned.

For example, Rabbi Moshe Tendler, a professor of biology and of Jewish medical ethics at Yeshiva University, has said that human cloning, in certain circumstances, could have moral and therapeutic uses. He told the PBS show *Religion and Ethics Newsweekly*, "Give me a circumstance such as a family who was murdered during the Holocaust, leaving but one survivor, a sterile male, I certainly would clone him." He went on to point out an example of how a cloned human being could serve therapeutic purposes as well:

> Likewise, show me someone whose life is being threatened and in need of a bone marrow or same cell transplant and there is no match available, I certainly would clone him and then use the resulting child as a source of same cells, a source of bone marrow, and this child would be doubly loved for himself or herself, and for saving the parent.

Rabbi Michael Broyde, a law professor at Emory University in Atlanta, writes in *Jewish Law* that Jewish law does not prohibit cloning, and that it can even be considered a mitzvah, or a good deed.

For example, if someone who is ill could be cloned to ensure a match in a bone marrow transplant, and if a transplant from the cloned being could save that person's life, he writes, "Jewish tradition might regard this procedure as involving two good deeds: having a child and saving a life."

BioFact

The Jewish tradition includes legends about a human-like being called a "golem" that in some ways may be considered similar to a clone. A golem is a being created from dirt or mud through mystical means. Legend has it that golems have been created over the past 600 years to help the Jewish people in times of need. It is also generally believed that the golem legend was influential in shaping Mary Shelley's book *Frankenstein*.

Judaic View of Stem Cell Cloning

As a general rule, most Jewish thinkers back therapeutic stem cell cloning. Jewish law does not say that a human life begins when a sperm fertilizes an egg, and according to a paper commissioned by the National Bioethics Advisory Commission, "A fetus has no status during the first 40 days. More to the point, an embryo existing outside of a woman has no legal status in the Jewish tradition. Therefore, there is no intrinsic objection to embryo research." From a sermon given by Rabbi Barry Block, "according to our tradition, a person becomes a person only upon birth."

Because Judaism does not consider an embryo a human being, there is nothing wrong with using embryonic cells to cure disease. Because there is such a strong Jewish tradition to heal, many Jewish thinkers strongly support therapeutic stem cell cloning. For example, Rabbi Tendler said at a panel, sponsored by the Pew Forum on Religion and Public Life, titled "Human Cloning: Religious Perspectives" that "stem cell research is the hope of mankind." He called a bill to ban stem cell research "an evil being perpetrated on America."

Other Religions and Cloning

Other religions, such as Islam, Hinduism and Buddhism, of course, have points of views about stem cell cloning. Those religions, much like Judaism, have no single, central authority, and so there is no single point of view they take.

BioDefinition

A **fatwa** is a legal opinion or a legal ruling issued by an Islamic scholar. An individual scholar can issue the ruling; it does not have to pass muster with a committee.

Islam and Cloning

Islam, for example, is made up of a multiplicity of branches and schools of thought, with widely varying points of view, so there is no way to generalize among them all.

The Cairo-based Muslim scholar Yusuf al-Qaradawi issued a *fatwa* against human cloning, saying it is "completely prohibited" because it "contradicts the diversity of creation." He also noted that it can help destroy family life and relationships.

However, his point of view is not unanimous. According to the ummahnews.com site, Lebanon's top Shiite scholar Ayatollah Mohammad Hussein Fadlallah told Tehran radio that cloning humans should be allowed if its positive impacts ultimately outweigh its negative impacts. He said that "cloning would have negative repercussions on the emotional, social and family arenas," but that "human cloning could have positive health aspects and help find new discoveries that might be used in treating chronic diseases." The balance between those two will ultimately decide whether cloning should be allowed, he said. But he said there is no religious reason to ban cloning.

BioFact

The Council on American-Islamic Relations did a survey in 2001 to find out what Muslims felt about cloning. Eighty-one percent of respondents said they were opposed to human cloning.

Buddhism and Cloning

It is even more difficult to pin down the Buddhist stand on cloning, and it is difficult to find where Buddhist scholars come down on the issue. Damien Keown, a professor at Goldsmiths College in London and an expert on possible Buddhist responses to cloning, told MSNBC that Buddhism doesn't have the kind of moral, fundamental opposition to cloning that Catholicism does. He claims, in the words of MSNBC, "Buddhism teaches that life can come into being through supernatural phenomenon like spontaneous generation." He told the cable channel, "Life can thus legitimately begin in more ways than one." In a paper, he noted that the "technique in itself would not be seen as problematic."

But because Buddhism practices respect for all living things, destroying cells can run contrary to Buddhism, and many cells have to be destroyed in cloning research, he notes.

Hinduism and Cloning

Hinduism does not take a stand on the issue of cloning, and there is no central authority, for that matter, who can take that stand. The influential journal *Hinduism Today*, however, offers an excellent overview of the issues that Hindus consider in the debate. The editorial was written after the editor interviewed Hindu thinkers about cloning.

The editorial noted:

> Most Hindu spiritual leaders we spoke to were less concerned for the moral issues … human cloning than for the practical need. Why do this? they asked. Will it help us to draw nearer to God if we have such bodies? Will the soul's evolution toward Self Realization be advanced one millimeter? Will the inner consciousness be enhanced?

The editorial noted that most thinkers were against it, because the answer was "no" to all those questions. But it said that some said that cloning could prove to be useful.

In summation, the article noted "If done with divine intent and consciousness, it may benefit; and if done in the service of selfishness, greed and power, it may bring severe karmic consequences."

BioFact

Hinduism includes many stories and legends concerning human beings being produced through asexual means somewhat similar to cloning. For example, for every drop of blood spilled by the mythological demon Raktabija ("blood drop"), another of him sprang into being. The Lord Ganesha was said to have been created from the skin of His Mother. And Lord Murugan was conceived by a spark from the god Shiva's third eye.

The Least You Need to Know

- The Catholic Church makes no distinction between human cloning and embryonic stem cell cloning, and would ban both techniques.
- Protestant denominations are largely opposed to human cloning. Some oppose it on religious grounds, and others because it would only help the privileged.
- Protestant denominations are split over whether embryonic stem cell cloning should be allowed. The most conservative groups would ban it, while centrist and liberal groups favor it.
- There is a great deal of disagreement in the Jewish community over human cloning, with some saying that it should be banned, and others saying that it should be allowed.
- Most Jewish thinkers support embryonic stem cell cloning.
- There has not been a great deal of debate in the Islamic, Hindu, and Buddhist communities about embryonic stem cell cloning.

Chapter 17

Beyond Cloning

In This Chapter

- Amazing advances in bioengineering and DNA research
- The creation of man-machine hybrids
- All about brain transplants
- Telomeres and immortal cells
- Ray Kurzweil's vision of the future

Cloning and genetics are probably the most exciting areas of medical research, but they're far from alone in attracting attention. Researchers are looking at ways that genetics-related research can be combined with other disciplines, such as nanotechnology, artificial prosthetics, computer technology, and more.

They're looking to help cure diseases, but for many, that is only a starting point. Some researchers see the day when man and machine will become inextricably connected to one another, when computer chips are embedded in our brains, when memory can be implanted via computer chips, and when we can change our DNA and approach immortality.

Yes, this is all very futuristic stuff. But as you'll see in this chapter, there are already technology and advances that point the way toward the future.

The Future Is Now: Amazing DNA and Bioengineering Advances

The great Irish poet William Butler Yeats wrote the following lines in 1927 in his poem *Sailing to Byzantium*:

> Once out of nature I shall never take
> My bodily form from any natural thing,
> But such a form as Grecian goldsmiths make
> Of hammered gold and gold enameling.

In these lines, Yeats expresses a universal longing, not only for immortality, but also of the apparently innate human desire to be something better than we are, to be purer or more finely wrought in some way—in short, to be better and more than human.

BioSource

For the full text of the poem *Sailing to Byzantium*, go to www.online-literature.com/yeats/781/. For an excellent edition of Yeats' collected poems, get *The Collected Poems of W. B. Yeats*, edited by Richard J. Finneran and published by Scribner.

Poets express the longings of the human heart; scientists, in a more down-to-earth way, try to satisfy them.

And so today, more than 75 years after Yeats wrote those lines, a variety of biotechnology, computer, and genetics technologies are coming together in sometimes unanticipated ways to point us toward if not immortality, at least longer life, and hybrids of man and machine that can make human life better.

BioDefinition

According to the World Transhumanist Association, **transhumanism** is: "The intellectual and cultural movement that affirms the possibility and desirability of fundamentally improving the human condition through applied reason, especially by developing and making widely available technologies to eliminate aging and greatly enhance human intellectual, physical, and psychological capacities."

These advances are mostly in the experimental or imaginary stages, and it's not even clear whether they will ever come to fruition. Taken by themselves, they may be small advances, but as a group they hold out compelling visions of genetically and mechanically advanced humans. In fact, there is already a movement afoot that promotes the engineering of the human body in any way possible if humans can be improved by that engineering. The movement is called *transhumanism*, and while at the moment it's far from a groundswell, it is gaining increasing adherents.

What kind of advances are we talking about? Consider some of these imaginary scenarios:

- Man-machine hybrids in which the functioning of the human body and brain is enhanced by the addition of high-speed computer chips and artificial chromosomes
- Human chromosomes manipulated to give people almost immortal life spans
- The "downloading" of memories from a dying human brain into a silicon-based device that can be implanted in a new body, in essence giving that new body the mind and memories of the dying person

Yes, these all sound far-fetched, and none are yet happening today. Yet, as you'll see in this chapter, work is being done on all of these scenarios. For example, as we go to press, scientists have plans to test an artificial brain part that could be the first step to downloading human memory onto a chip and then implanting that chip in a brain. Scientists already use artificial devices implanted in brains to help those brains function in patients with Parkinson's disease. And they have identified an "immortal enzyme" and gene that may help cells live forever.

BioSource

If you're interested in transhumanism, check out these websites and organizations: For the World Transhumanist Organization, go to www.transhumanism.org. For information, articles, and news about transhumanism, go to www.betterhumans.com.

Yeats said in his poem *Sailing to Byzantium*, "Gather me into the artifice of eternity." In the rest of this chapter, we'll take a look at some of that real-life and future artifice.

A Look at Brain Implants

The idea sounds as far-fetched as humanly possible: Brain implants of silicon-based computer chips that work seamlessly with the brain, enhancing all of its capabilities. Or implants of literal memory chips—computer chips that have a person's entire memory and personality embedded in them, so that they can be implanted in a new brain as a way to achieve immortality.

Far-fetched, yes. Impossible, maybe not.

In fact, human brain implants are already in use, and have been approved by the Food and Drug Administration (FDA) to treat Parkinson's disease. A great deal of work has gone on with animal brain implants. Researchers have already programmed a mathematical model of memory and placed it onto a computer chip, which they expect to implant into the brain. So we're already well on our way to sophisticated brain implants.

BioFact

A great deal of research is going on into brain/computer interfaces (BCI) that may eventually allow direct communications between computers and the brain. Researchers at Emory University have developed implantable electrodes that can fuse with brain cells. The electrodes are tiny glass cones with holes inside them, and inside the cones are microscopically thin gold wires, electrodes, nerve tissue taken from the leg, and "trophic factors" that induce brain cells to grow into the cone.

Brain Implants and Parkinson's Disease

Parkinson's disease affects an estimated 1.5 million Americans. The disease causes tremors, rigidity, difficulty in moving, and, in some people, intellectual deterioration among other symptoms.

A brain implant approved by the FDA, however, can help control the tremors associated with the disease. Approved by the FDA, the Activa Parkinson's Control System is made up of electrodes that are implanted into the brain and are connected to wires under the skin to a pulse generator that is implanted in the chest or the abdomen in a control box. This pulse generator sends a steady stream of electrical impulses to the brain, which cuts down or eliminates the tremors. An external computer communicates with the control box via radio waves and can control the voltage and pulse of the electricity, as well as which wires should carry the current at any given time.

BioFact

The same device used to control tremors in Parkinson's patients has also been used to control tremors in people who have cerebral palsy.

Using Memory Implants and More

Work is already taking place on memory implants that could help restore memory to a brain damaged by stroke, epilepsy, or dementia. Ultimately, some say, memory could theoretically be downloaded into a memory implant, and that implant could then be placed in a different brain.

Researchers at the University of Southern California have built a mathematical model of a part of the brain called the hippocampus, which works to store memories—and they've stored that model onto a computer chip. They're building a model not of a human hippocampus, but of a rat's. They did it by slicing sections of the hippocampus, electrically stimulating the slices, and seeing what types of electrical outputs resulted. Eventually, they built a model detailing the circuits, and programmed that onto a chip.

They plan to first test the chip on rats, and then move on to monkeys, after building a computer model and chip of a monkey hippocampus. They will attach the chips to brains via electrodes, stop the animal's own hippocampus from working, and then see whether the chip can perform the functions of the hippocampus, including processing memory. If the animal is able to store new memories while attached to the chip, the experiment will be a success—and humans, ultimately, may be next.

BioFact

The next time you hear someone call a piece of art "ratty," it may have a completely different meaning than you think. Several thousand rat brain cells in a Petri dish at the Georgia Institute of Technology are creating drawings thousands of miles away at the University of Western Australia in Perth, Australia. The neurons are connected to electrodes that pick up the signals and transmit them to a computer that translates them into the movement of three colored markers posed above a white canvas. The signals are sent over the Internet to the University of Western Australia, where the actual painting occurs.

The hippocampus-on-a-chip is only one of the experiments going on in a field known as neural prosthetics. At Johns Hopkins University, work is going on to create an array of photosensors that could be fitted onto the back of a damaged human retina, allowing the blind or partially-sighted to see.

Immortal DNA and Telomeres

The average life span of someone in the United States today is in the 80 year range, far above what it was a century or more ago. But even that isn't enough for many of us. We want far longer life spans, and maybe even immortality.

DNA research may hold the key to longer lives, and in the fantasies of some, to immortality as well.

One of those keys is a molecular chain of repeating DNA segments called the telomere, which caps the ends of chromosomes. Telomeres appear to protect the interior gene-containing portion of the chromosomes so that they are not lost, damaged, or scrambled, and may aid in the copying of chromosomes when cells divide. Without telomeres, cells die.

Every time a cell divides, its telomeres shorten slightly. Ultimately, the telomeres become so short that they can no longer function properly, and the cell dies.

BioFact

Telomere shortening may be one of the reasons that many cloned animals, such as Dolly the Sheep, appear to suffer from disease such as arthritis related to premature aging. Because the DNA of a cloned animal is taken from an already grown animal, its telomeres have already been shortened. Because of that, its cells may not live as long, and so the animals may die of premature aging. For more information, see Chapter 6.

Scientists have discovered an enzyme, telomerase, that appears to repair and lengthen the telomeres in human cells. Normally, telomerase is only used for the formation of sperm and maybe egg cells, and in the creation of stem cells so that those cells contain "young" telomeres that are not shortened. The gene that creates telomerase is called human Telomerase Reverse Transcriptase (hTRT), and except for the formation of sperm and maybe egg cells and in stem cells, the gene is normally "turned off."

There is a link between telomerase and cancer, however. Many cancer cells have the hTRT gene turned on, which then produces telomerase so that the cancer cells' telomeres do not shrink. This means that those cancer cells are "immortal" and so they do not die as do normal cells—they continually replicate themselves uncontrollably without dying.

Researchers hope that telomerase and the hTRT gene can alleviate the effects of human aging. If there were a way to release telomerase, but not have it lead to cancer, the thinking goes, then perhaps the deleterious effects of aging could be avoided. According to the website of the Shay/Wright laboratory at the University of Texas Southwestern Medical Center, "In the future, it may be possible to take a person's own cells, manipulate and rejuvenate them without using up their life span and then give them back to the patient."

BioFact

Some cancer researchers believe that the hTRT gene may be a key to helping cure cancer. If researchers can find a way to turn off the gene in cancer cells, then the cancer cells will no longer produce telomerase and be "immortal" and so the cancer cells will not proliferate out of control.

BioSource

There's an organization for everything, and so it's no surprise that there's one for those who believe that it is possible to make humans immortal. If you're interested, visit the website of the Immortality Institute at www.imminst.org.

Ray Kurzweil's Cyborg Vision

Ray Kurzweil has long been one of the premier computer scientists, with an impressive record of inventions. He developed a new way of performing optical character recognition, a new kind of flat-bed scanner, and the first full text-to-speech synthesizer. He combined the three technologies and built the first print-to-speech reading machine for the blind, the Kurzweil Reading Machine. With Stevie Wonder as a musical advisor, he built the Kurzweil 250, the first computer-based instrument that could recreate the musical sound of the piano and other orchestral instruments. He's had other inventions as well, and won numerous awards, including the National Medal of Technology, the nation's highest honor in technology, and Inventor of the Year by MIT and the Boston Museum of Science.

When Kurzweil speaks, people listen.

Kurzweil believes that ultimately genetics, *nanotechnology* bioengineering, and computer technology will be used to enhance the human body and mind in ways not yet imaginable. He envisions brain-embedded nanorobots that can augment all brain functions, allowing us to think faster and more efficiently, record and play back feelings, thoughts and dreams at will, and switch instantly between virtual reality and sensory reality.

BioDefinition

Nanotechnology is the research and development of very small devices and machines that are measured in the billionth of inches.

The core of our being will become mechanical—he says that the very nuclei of our cells will be replaced with man-built, nanotechnology structures with all the genetic materials required to make the proteins necessary for life. If we can build those structures, we can alter them, which means we can change our biological structures at will. Although this is still the realm of science fiction, researchers have already created artificial chromosomes, implanted them into mice, and those chromosomes have worked and been passed down to future generations. (For more information see Chapter 21.)

In fact, Kurzweil says, perhaps the nonbiological parts of the body will eventually overwhelm the biological parts of the body. Those biological parts may eventually no longer be needed. There is the possibility that to be human will eventually mean to be made largely from machine parts.

The Ted Williams Story

In the Boston area, where Preston lives, the slugger Ted Williams is treated as a baseball god. He was the last batter to hit .400 over a full season, had a lifetime batting average of an astonishing .344, and had 521 career home runs. There are many who believe he was the best hitter who ever lived.

Williams died in 2002, and his after-death fate has elements of farce, the grotesque, tragedy, and seaminess—and it's an example of what can happen when bioengineering goes wrong.

BioDefinition

Cryogenics is the practice of freezing the body of a person who recently died, as a way to preserve it so that it may be resuscitated in the future when a cure has been found for the disease that killed the person.

After he died Williams' body became involved in a tug-of-war between his children. On one side was his son John-Henry and daughter Claudia, and on the other side his daughter Barbara Joyce. John-Henry, claimed that his father wanted to be cryogenically frozen and his DNA preserved, in case at some future time he could be cloned from his DNA, or there would be a way to bring his frozen body back to life when a cure for the disease that killed him is found.

Shortly after Williams died, his son John-Henry had the Alcor Life Extension Foundation rush the body away so that it could be frozen in liquid nitrogen at subzero temperatures. Legal battles ensued, with other relatives appalled, but ultimately John-Henry won and his wishes are being followed.

Unfortunately, though, according to *Sports Illustrated* and Larry Johnson, the former chief operating officer of Alcor who was there when Williams was preserved, Williams' body was treated in a horrifyingly grotesque manner. According to the magazine, shortly after Williams died, his body was flown to the Alcor Arizona facility and that evening his head was removed from his body and stored in a liquid nitrogen two-and-a-half-foot-high cylindrical steel container. His body was frozen as well and stored separately from his head.

There were problems storing Williams' head, said the magazine. Two holes were drilled in it so that the brain condition could be observed, but then a "huge crack" formed in the head, and after nine other cracks appeared, the head was removed from its container and placed in a special "neuro-can."

The magazine also reported that silver packets floating in the liquid nitrogen above Williams' head contained samples of William's tissue that were going to be harvested for his DNA. These were missing, said that magazine, which added that John-Henry planned to sell the DNA samples to the highest bidder.

Making the story even more of a farce is that John-Henry owed Alcor $110,000 for the storage, says the magazine. According to the magazine, an Alcor board member and advisor joked that as a way to convince John-Henry to pay up, they should throw away the body, send it in a "frosted cardboard box" C.O.D. to John-Henry, or else post it on eBay as an auction item.

As we go to press, the story continues with no resolution. It's a cautionary tale about the desire for immortality via genetics and engineering.

BioSource

The Alcor Life Foundation claims to be "the world's largest provider of cryogenics services." On its website the foundation claims, "With the many advances in modern science, such as: DNA mapping, stem cell research, therapeutic cloning, human genome studies, and the emerging discipline of nanotechnology, the possibility of living a longer, more productive life is becoming more realistic with each passing day." For information about the organization, go to www.alcor.org/.

The Least You Need to Know

- Transhumanists believe that we will be able to create man-machine hybrids that will combine computer technology, nanotechnology, and genetic technology.
- Some people believe that we may some day be able to "download" our memories into computer chips and implant them in brains.
- Scientists have created a chip that contains the memory-creating portion of a rat's brain and are experimenting to see whether that chip can replace normal brain functioning.
- Electrodes implanted into human brains can help stop the tremors caused by Parkinson's disease and cerebral palsy.
- Some people believe that telomeres, which aid in the protection of DNA and replication of chromosomes, may hold a key to longer human life.

Part 4 Bioengineering, Genetics, and the Future

Bioengineering—the ability to engineer plant and animal life by manipulating their genes—gives us more power over all living things than our ancestors ever imagined. We can combine genes from plants with other plants, from animals with other animals, and from plants with animals. We can engineer salmon so they grow more quickly, and plants so they have longer growing seasons. To a certain extent, we can even bioengineer people.

This section looks at some of the thorniest issues of our times—given that we have untold power over how to shape life, how should we use that power? In fact, should we even use it at all? Those are some of the issues we'll explore in this part, and we'll take a look toward the future as well.

Chapter 18

Bioengineering—What Is It and How Does It Work?

In This Chapter

- Bioengineering explained
- How plants and animals are bioengineered
- The benefits of bioengineering
- Can terrorists engineer bioweapons?
- How bioengineering can create "designer drugs"

Mankind has long excelled at massive engineering projects. In modern times, think of dams such as the Hoover Dam and the Aswan Dam. In ancient times, think of the pyramids and Stonehenge.

Today, engineering projects are getting small—exceedingly small. The frontiers of engineering are no longer associated with the big and visible. Instead, they are taking place at the microscopic level—inside the chromosomes of plants and animals. And the effects of this bioengineering will be no less dramatic than the effects of massive engineering projects of the past.

In this chapter, we'll look at what bioengineering is, how it works, and how it is being used today.

It's Not About Bridges: What Is Bioengineering?

Bioengineering is one of those terms that most of us have heard about, but most of us don't quite fully understand what it means. Does it have something to do with building holding ponds for fish farms? Building new bodies in a machine shop?

BioDefinition

If you're looking for the Grade A, government-approved definition of bioengineering, what better place to turn than the National Institutes of Health (NIH), the premiere government arm that funds medical and biological research? Here's the definition of the term decided upon by the NIH Bioengineering Consortium. Be forewarned; it's a mouthful: "**Bioengineering** integrates physical, chemical, or mathematical sciences and engineering principles for the study of biology, medicine, behavior, or health. It advances fundamental concepts, creates knowledge for the molecular to the organ systems levels, and develops innovative biologics, materials, processes, implants, devices, and informatics approaches for the prevention, diagnosis, and treatment of disease, for patient rehabilitation, and for improving health."

It's enough of a vague term so that it means different things to different people. To engineers, it means applying the principles of mechanical engineering to the creation of biomechanical systems—for example, attempting to build a human heart or another organ from either nonbiological materials, or a combination of nonbiological and biological materials.

However, the broad definition takes in a whole lot more ground. It includes a remarkable number of different disciplines, including genetic engineering, nanotechnology, biomechanics, biomaterials, and a whole lot more. In fact, some people apply the term even more broadly, and use it to cover environmental engineering, such as for erosion control, stream restoration, and wetland restoration. But most people don't use the term to refer to that.

Many major universities include bioengineering departments these days, from Berkeley to Clemson, University of Pittsburg, Penn State University, and beyond.

However, for the purposes of this book, we're going to talk about bioengineering in a more limited sense. We'll cover only those aspects of bioengineering that relate to cloning and genetic engineering.

BioFact

Building artificial organs is not merely a futuristic vision—it's being tested today. These artificial organs may be a combination of biological and nonbiological materials. For example, artificial livers have been tested for several years. In some of the most recent tests, a device called *Extracorporeal Liver Assist Device* (ELAD) is being tested at a dozen medical centers across the country. The ELAD is a large device the size of a washing machine that is hooked up to a patient. It includes filters filled with cultured human liver cells. Those cells are able to perform liver functions such as manufacturing certain proteins and removing some toxins from the blood. It is used temporarily to keep someone alive while waiting for a liver transplant.

Specifically, we'll look at the following:

- Using bioengineering to create new plants and animals
- Using bioengineering to bioengineer babies
- The role that biotech companies play in bioengineering and cloning

BioSource

To access the federal government's central information point about bioengineering, go to the National Institutes of Health Bioengineering Consortium website at www.becon.nih.gov/becon.htm.

We'll cover those in much more detail in Chapters 19 through 24. In this chapter, we'll give a brief overview of how genetic bioengineering works, and take a look at issues such as bioengineering and terrorism, and so-called "designer drugs" that may be able to be created using genetic bioengineering techniques.

How Genetic Bioengineering Works

Genetic bioengineering enables researchers to create plants and animals never found before in nature by manipulating their genetic material. We're not talking about creating tigers with wings and elephant tusks, or pine trees with year-round flowering orchids. Rather, the genetic manipulation is much more subtle and often the results of it are not visible. A genetically engineered ear of corn, for example, usually looks like its nongenetically engineered cousin; and a genetically engineered mouse can look just like a normal mouse.

Usually a single gene, or a small group of genes are altered, added, or stopped from functioning. The purpose is not to create circus freaks. Rather, the purpose is much more limited. Remember back to our initial chapters about DNA, that a gene is essentially a blueprint for creating a protein or a family of proteins? With genetic engineering, the purpose is usually to insert a gene into a plant or animal so that plant or animal creates a protein that it normally wouldn't. Or the purpose may be to *knock out* a gene—that is, stop it from making the protein that it would normally make.

BioDefinition

To **knock out** a gene means to change it so that it cannot produce the functioning protein it was programmed to produce.

Stated that way, genetic engineering may not sound particularly exciting. But remember, proteins are the key to life—they help build all living things and keep them functioning. If you can manipulate the proteins that a plant or animal makes, you can alter many of its basic functions. In addition, by manipulating their genes, you can create animals and plants that are more useful to human beings.

All this may sound rather abstract, so let's take just a few instances. All of the following are either already in production or have been tested. More details about these examples can be found later in Part 4.

- Bioengineered plants, including a common one called Bt corn, can produce their own natural pesticides that are safe for humans and other animals enabling pesticide use to be cut dramatically.
- Bioengineered plants can be engineered to be drought-resistant, withstand the cold, or have longer growing seasons than normal plants, helping feed the hungry.
- Bioengineered plants can be engineered to have extra nutrients that they don't normally have, providing extra nutrition to the poor in countries in which many people suffer from malnutrition. For example, so-called "golden rice" contains elevated amounts of Vitamin A, and could contribute to preventing blindness in children who suffer from Vitamin A deficiency.
- Bioengineered plants can produce some industrial biochemicals at lower cost and with less damage to the environment than when those biochemicals are produced by factories.

BioWarning

There are some people who warn that bioengineering plants and animals can be dangerous and can harm human health and the environment. For more details, see Chapters 20 and 23.

- Bioengineered plants and animals can produce drugs and other pharmaceutical biochemicals—drugs and biochemicals that are too expensive to produce in a more traditional manner, or that can be produced at lower cost through bioengineered animals.
- Bioengineered animals can be used to grow organs for human transplantation.

As you'll see later in this chapter, genetic bioengineering has other uses as well, such as genetically engineering viruses to fight cancer and building designer drugs.

Bioengineering Plants

The techniques vary somewhat according to whether you're bioengineering a plant or an animal. And there are many different ways to bioengineer each. We'll cover the techniques in only the most general way.

Let's start with bioengineering a plant. Say you wanted to bioengineer a plant to produce a natural pesticide. First, you identify the pesticide you want the plant to produce. Next, you identify the gene or genes in a specific plant or animal that produces that pesticide. For example, the Bacillus thuringiensis (Bt) bacterium has a gene that creates the Bt protein that kills some insects but is harmless to people and animals.

You isolate that gene, by splicing it into something called a *bacterial plasmid*, a small ring of DNA. You then insert that plasmid into a soil bacteria, and mix it with plant cells. The gene for producing the pesticide becomes part of the chromosomes in a small number of plant cells. Those cells are then grown into whole plants. Those plants will now have the gene for producing the natural pesticide and can produce that pesticide when they grow.

BioDefinition

A **bacterial plasmid** is a circular piece of DNA in bacteria that copies itself when the cell divides. The Ti bacterial plasmid can cause tumors in plants but not in animals. The Ti plasmid is frequently used in plant genetic engineering because it naturally invades plant cells and inserts its DNA into the DNA of the cell.

Bioengineering Animals

Bioengineering animals is somewhat different. The first steps are the same—identifying the gene you want to implant in an animal and then isolating that gene.

From there, though, things differ. You need to insert the gene into the embryo of an animal, so when the animal grows up, the gene will work. It's not as likely that the gene will work if you inject it into an already grown animal and it is much more difficult to do as well.

There are a variety of ways to insert the gene into the embryo. One common way is via animal cloning. Using this technique the new gene is first inserted into the DNA of an animal cell, not into an embryo. The DNA of the animal cell with the extra cloned gene then is then inserted into an egg whose nucleus has been removed. The egg, with the cloned DNA, develops into an embryo and then into a fully grown animal. When the animal becomes an adult, the gene does its work.

Genetically Modifying Microbes to Cure Disease

Genetic bioengineering doesn't have to involve large plants and animals—in fact, it can also be applied to microbes and even to viruses. Viruses are essentially just bits of genetic material wrapped inside proteins and they can be genetically manipulated as well.

Viruses in particular may hold the key to helping cure diseases. As explained earlier in this chapter, sometimes they are used to help genetically engineer other plants and animals. Because of their unique abilities to enter cells and change cellular DNA, or take other actions on cells, genetically engineered viruses hold a great deal of potential.

A Cure for Glioma?

Let's take an example of how genetically engineered viruses may help kill a specific form of cancer. Glioma is an especially dangerous form of brain cancer that usually recurs even after chemotherapy, radiation treatment, and surgery. People who get the disease usually die within a year.

BioFact

Researchers have genetically engineered viruses to kill gliomas, including the virus that causes polio. The polio virus kills only glioma cells in the brain because it binds to a unique protein present on the surface of glioma cells. However, injecting the polio virus can also infect normal nerve cells and cause polio. Scientists altered the virus so that it could not infect nerve cells by replacing a snippet of genetic RNA with a similar RNA from a rhinovirus. A rhinovirus is a common kind of virus that causes the human cold. The bioengineered polio virus killed glioma cells and did not cause polio because its ability to infect nerve cells was destroyed. The bioengineered virus has been tested on mice, but not yet on human beings.

Gliomas grow uncontrollably in part because their cells have a mutated gene that produces a defective Rb protein. Mature cells divide slowly because their Rb protein regulates the replication of chromosomes, and so a crippled Rb protein can lead to uncontrollable cell growth and cancer.

Researchers have genetically engineered a virus called Delta-24-RGD so that it can enter and destroy the cells of the glioma tumor but leaves the rest of the body's cells alone. The virus was genetically engineered so that it can only replicate in cells with crippled Rb—the normal Rb protein will kill it. Additionally, the virus was engineered so that it could more easily attach to glioma cells so that it could insert itself into the cells and kill them.

As of this writing, tests have been conducted on mice and the results were encouraging enough so that human trials are expected to be conducted.

Genetic Engineering and Interferon

Interferon is a cancer-fighting drug that is naturally produced by the human body when it is attacked by a virus. It was first discovered in 1957 and was studied for years. Although its usefulness was immediately recognized, it could not be used in patients because of how difficult and expensive it was to obtain. It is produced in such small amounts that the blood of 90,000 donors would only be able to provide three one-hundredths of an ounce of interferon, and that interferon would be only about one percent pure. In 1978, for example, just one dose of impure interferon cost $50,000.

Genetic engineering has changed all that. The human gene for producing interferon was cloned into a bacterial plasmid and then implanted into bacteria that can be grown in cultures producing trillions of copies of the gene. The bacteria were able to mass produce interferon so that by the middle of the 1980s, it was being produced for about one dollar a dose.

Bioengineering and Terrorism

If you want to know what keeps some anti-terrorism experts awake at night, consider this: Deadly diseases can be bioengineered so that they are even *more* deadly and more easily spread by terrorists.

That's not an imaginary horror. The former Soviet Union, and most likely other countries as well, have genetically bioengineered viruses and bacteria so that they make more efficient weapons.

According to the *Chicago Tribune*:

> "The Soviets were the first to attempt to bioengineer natural pathogens into more virulent ones when, in the 1980s, they secretly produced anthrax spores to be resistant to some antibiotics. They also unsuccessfully tried to combine smallpox with the Ebola virus to make a hardier killer."

If there is a more horrifying thought than a disease that combines smallpox with the Ebola virus, we can't think of it.

CAUTION

BioWarning

The Ebola virus is one of the most fearsome viruses on earth. It causes Ebola hemorrhagic fever, which kills from 50 to 90 percent of all people who get it. Those infected first have a sudden onset of fever, weakness, muscle pain, headache and a sore throat. As the disease develops, it is followed by vomiting, diarrhea, severe kidney and liver damage, and extreme internal and external bleeding.

Those warning of the possibility of bioengineering being used to create biological weapons of mass destruction are not at the fringes of scientific thought—they are directly in the center. Barry Bloom, dean of the Harvard School of Public Health told the *Tribune*:

> "Many of the genomes for agents that could be used in biological warfare are now matters of public knowledge. They are available to anybody in the world. If they want to mutate genes, or splice them and take them out, all they need is a couple of DNA primers, some DNA and a little skill, and that can be done."

Bioengineered weapons would be created in the same way that other bioengineering is done, by mixing and matching genes from different bacteria and viruses to produce a more deadly killer, or one that could be more easily spread.

Bioterrorism need not be done via killer diseases such as anthrax that target humans. Another potential target is agriculture and the food supply—bioengineered pathogens could be turned loose to kill plants and cause famine.

BioSource

For information about what the federal Centers For Disease Control and Prevention recommends be done in response to various forms of bioterrorism, go to www.bt.cdc.gov/.

Of course, the same techniques used to create bioweapons can also be used to counter them. For example, if terrorists create an antibiotic-resistant germ, scientists can figure out other ways to kill the germ or make people immune to it. Researchers have discovered a gene that makes mice resistant to anthrax and they could potentially use that knowledge to make more effective vaccines and drugs.

Building Designer Drugs

Researchers believe that ultimately biotechnology will lead to the creation of "designer" drug treatments—drugs and treatments that are specifically created to match an individual's genetic makeup and specific disease. These drugs could possibly include one-of-a-kind proteins that would be produced by genetic engineering, and would be based on the genetic makeup of individuals.

Scientists are already creating protein chips intended to reveal the exact mix of proteins in a patient's blood. Many diseases like cancer lead to the shedding of specific mixes of proteins into the blood. These chips should be able to quickly tell physicians exactly what type of disease is present. They can also be used to follow the disappearance of the disease as it is treated and predict whether it is reappearing in later life. There are also gene chips being developed that can suggest different treatments depending on someone's individual gene sequences. Eventually drugs may be created that would be tailored to the exact mix of proteins and genes in an individual.

Some designer drugs are already hitting the market or are being tested. These drugs are not designed for a particular individual, but are still designer drugs because they are aimed at blocking or changing specific biological reactions that occur with certain diseases, and curing the disease in that way. For example, traditional cancer chemotherapy or radiation treatment may kill many cells in the body, not just cancer cells. The first generation of designer drugs do not kill cells indiscriminately, and instead attempt to kill only tumor cells or block specific biological reactions within tumor cells.

One example of such a designer drug is Herceptin, developed by scientists at UCLA and the biotech company Genentech. The drug is designed to treat the large number of people whose breast cancer is caused by a mutation of the HER2 gene. The HER2 gene creates the HER2 protein that sits on the tumor cell surface and tells the cells to divide. The mutation causes too many copies of the HER2 gene. Too much HER2 protein is produced, which can lead to fast-growing, aggressive breast cancer.

Herceptin targets the cancer cells created as a result of this. It binds to the HER2 proteins on the tumor cell surface and prevents them from receiving signals that tell the cancer cell to divide. It also attracts immune system cells called natural killer cells that kill the cancer cell. In this way, it's a kind of scout that cuts the lines of communication to the tumor cells and then tells the immune system to kill them.

BioSource

For more information about Herceptin, go to www.herceptin.com.

The Least You Need to Know

- There are many definitions of bioengineering, but it often refers to the creation and manipulation of plants and animals using genetic engineering techniques.
- You bioengineer a plant or animal by inserting a gene into it from another plant or animal, or by disabling or altering one or more of its genes.
- Bioengineered plants and animals can offer many benefits, including creating low-cost pharmaceuticals and feeding the hungry.
- Countries such as the former Soviet Union bioengineered disease-causing viruses and bacteria to make them more lethal and to make them easier to use as weapons of mass destruction. Some people worry that terrorists may ultimately do the same thing, and then use those bioweapons.
- Most researchers believe that "designer" drugs and treatments will be created that can be engineered for people's specific genetic makeup and disease.

Chapter 19

Pig-Men and Fish-Tomatoes

In This Chapter

- A definition of transgenic plants and animals
- How transgenic animals are created
- How transgenic plants are created
- Common transgenic plants and animals
- The benefits of transgenic plants and animals

"A rose by any other name would smell as sweet," Shakespeare wrote hundreds of years ago. But in his day, of course, a rose was a rose, was a rose.

Today that might not be the case. A rose might have a salmon gene implanted in it, or a chicken might have a rose gene inside it. You can't even tell by looking, touching, or smelling, for that matter.

Increasingly common today are transgenic plants and animals in which one plant or animal may contain genes from another for a particular purpose. For example, corn may be given a gene that will create a natural insecticide, or sheep may be given a gene that will have them secrete medicine in their milk.

In this chapter, we'll look at what transgenic plants and animals are, how they're created, and take a brief look at their benefits.

What Are Transgenic Plants and Animals?

"Johnny! Time for dinner!" Mom calls upstairs to her freckled, sweet-faced ten-year-old son.

Johnny comes tearing down the stairs to the dinner table.

"My favorites!" he says, staring at the bounteous spread before him. "Corn genetically engineered to produce natural pesticides; tomatoes implanted with Arctic fish genes; salmon bioengineered to grow faster by producing high amounts of growth hormones; golden rice that uses genes from daffodils so it has more Vitamin A; scalloped potatoes with chicken genes; cornbread with a firefly gene … Yum! Let's dig in!"

No, that's not a scene from the present, but give it a few years—it may be one in the future. And, in fact, all of Johnny's favorite foods at the dinner table either are already in production or are being tested as we speak. In fact, 68 percent of all soybeans grown in the United States were genetically modified in 2001 and 69 percent of all corn grown in the United States in 2001 was genetically modified. The numbers for the rest of the world were not nearly so high, but they're increasing.

And that's just talking about food. Plants and animals are being genetically modified for many other reasons as well—to produce pharmaceutical chemicals and industrial chemicals, to aid in medical research, and possibly even to create pigs that can be used to donate their organs to people who cannot get organs otherwise.

These plants and animals are called *transgenic*. No, that doesn't mean an animal that has had an operation to change its gender from male to female or vice versa. A transgenic plant or animal is one that has had its genetic makeup altered by having a gene from another plant or animal implanted into it, or by having one or more of its genes "knocked out" so that it does not function properly.

BioDefinition

A **transgenic** plant or animal is one that has been given a gene from another plant or animal, or has had one or more of its genes "knocked out" so that it does not function properly.

Once a remarkable feat of science, today transgenic plants and animals are commonplace. For example, Bt corn has been given a gene from the bacterium Bacillus thuringiensis, commonly called Bt, which creates an insecticidal protein. The corn produces its own natural insecticide that kills the European corn borer, the most damaging insect pest in the United States and Canada. Vast amounts of Bt corn are grown in the United States each year.

Many other kinds of transgenic plants have been created as well. And transgenic animals, while not as commonplace, are starting to become more widely used. For example, the so-called "Harvard Mouse," is a transgenic mouse that was specifically created to develop cancer. Its MYC gene, which encodes proteins that regulate cell growth and cell differentiation, was mutated. Because the mutated myc proteins can't function properly, the mouse develops cancer when some cells grow wildly out of control. The mouse is used in cancer research—potential anti-cancer drugs are used on it to see whether they work.

Transgenic plants have long moved out of the laboratory and into the fields and elsewhere as Bt corn and other transgenic crops show. Animals are following as well, although transgenic plants are far more common—there is a good chance that you've eaten transgenic foods recently. We live in a world where genetic engineering and biotechnology have supplemented the traditional means of breeding to create new plants and animals. Depending on your point of view, genetic engineering and biotechnology are creating transgenic plants and animals that will help cure disease, clean the environment, and help us all lead better, longer lives—or else it's leading to a world headed down the path to destruction. No matter which you believe, the technology is here to stay.

How to Create a Transgenic Plant

Transgenic plants and animals are created using similar methods, although there are some variations between them. We'll start off by looking at how to bioengineer plants, because those techniques have been around longer and are in more widespread use.

BioFact

When you think of tobacco, you may well think of cancer, because of the cancer-causing capabilities of the plant. But scientists are working on transgenic tobacco plants that may be able to produce a protein that prevents women from developing cervical cancer. Scientists at North Carolina State University have been working on inserting a gene into tobacco plants that produces a protein that is being tested as a vaccine against the papilloma virus, which can cause cervical cancer. They are doing tests to see whether the tobacco plants, when grown as crops, can produce sufficient amounts of the protein to make the vaccine affordable.

The bioengineering techniques for producing transgenic plants have become almost commonplace by now. The following figure shows how one would implant a gene for resistance to herbicides into a plant. First, a gene that produces a protein that will make a plant resistant to herbicides is identified and isolated. That gene is then spliced into a bacterial plasmid that also contains a marker gene. (A plasmid is a small ring of DNA.) That plasmid is then inserted into a kind of soil bacteria, which is mixed with plant cells. The plasmid with the gene for herbicide resistance and its marker gene then becomes part of the chromosomes in a few plant cells. The cells are then nourished in media that kills cells without the marker but lets cells with the plasmid grow. Because the plasmid also contains the gene for herbicide resistance, the adult plants will be resistant to herbicides.

The four steps to creating a transgenic, herbicide-resistant plant.

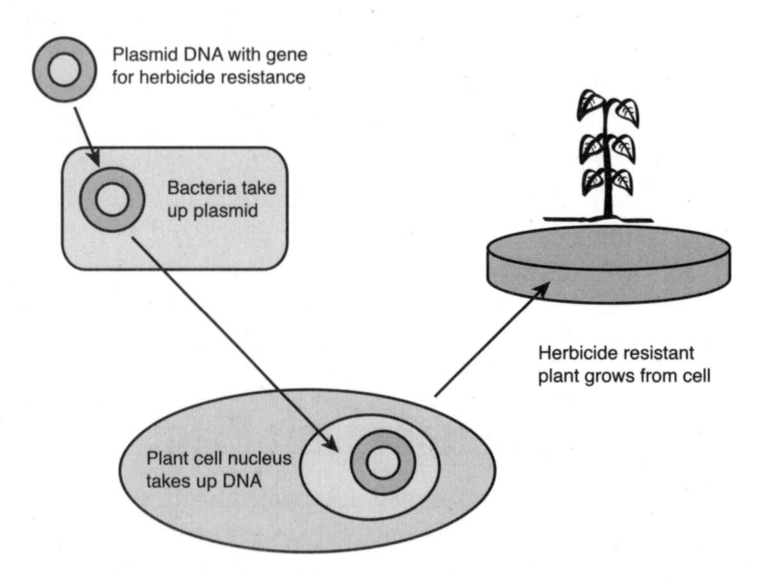
Plasmid DNA with gene
for herbicide resistance
Bacteria take
up plasmid
Herbicide resistant
plant grows from cell
Plant cell nucleus
takes up DNA

This isn't the only way to bioengineer plants, though. Harmless viruses can be used instead of bacteria, for example. However, there are other engineering techniques that can be used that don't require any kind of biological agent to get the gene into the plant. All of these use the same *vectors* to bring in the gene:

BioDefinition

A **vector** is a biological agent of some kind—for example, a harmless virus or plasmid—that is used to transfer a gene into a plant or animal cell.

- **Microinjection** The gene can be directly injected into a recipient cell, using a fine-tipped glass needle. After the gene is injected, it sometimes gets taken up by the plant's chromosomes. This only works with certain plant cells.
- **Poration** The target cells can be treated electrically or chemically to create pores in the cell membrane. The genes can then make their way into the cell via the pores and are sometimes incorporated into the plant's chromosomes in the nucleus.

- **Bioballistics** A "gene gun" can be used to fire genes into cells. This technique uses pellets or slivers coated with the DNA that is to be implanted into a cell. The gun fires the pellets or slivers, but the pellets or slivers never actually leave the gun and are stopped by a perforated metal plate before they can exit. But the speed of the pellets and the impact against the metal plate send the genes through the gun at high speed. They are able to penetrate into cells, where they can be taken up by the cells' chromosomes.

Note: All of these techniques are very inefficient and only a few cells take up the new gene. So how do scientists know which cells take up new genes and which don't? They attach a "marker" gene that is much easier to detect when both genes are taken up together by the plant cell. That way, it's easy for them to know which cells have taken up the new gene.

BioFact

Transgenic fish may be coming soon to a fish tank near you. The Taiwanese company Taikong Corp. has developed a commercially available pet fish that glows fluorescent green. The company spent $2.9 million to develop the pet fish by implanting a gene from a jellyfish into a zebra fish. It expects to create and sell 100,000 of the glowing fish every month. They sell in Asia for the equivalent of $17.40 each, and as you read this may have made their way to American shores—or fish tanks.

How to Create a Transgenic Animal

Similar techniques are used to create transgenic animals.

First, you need to determine what gene you want to implant into an animal. If you want the animal to produce a specific protein in its milk—for example, a blood-clotting protein—you need to find out what gene codes for that protein, and then isolate it outside the human body.

When you have the gene, you need some way to insert it into an embryo. You don't insert the gene into a grown animal, because there's no way to get it into all the proper cells in the proper organs so that the gene is turned on and producing the protein. For example, if you want the gene to produce blood-clotting protein in an animal's milk, you'll need to make sure that the animal's mammary cells contain the gene. You can't do that in a grown animal, so you instead insert the gene in an embryo. That way, the gene will be in all the cells in the growing embryo, and ultimately all the cells in the grown animal. You can ensure that it will be in the cells in the proper organ for the gene to express itself and create the protein.

But how do you get the gene into the embryo? There are a variety of different ways. Here are some of the most common:

- **Microinjection, poration, and bioballistics** These techniques are the same as outlined earlier in this chapter for creating transgenic plants. As explained, they don't involve bacteria or viruses.
- **Retrovirus vectors** *Retroviruses* are kinds of viruses that can insert DNA directly into chromosomes. The HIV virus is a kind of retrovirus, and in fact, the HIV virus has been bioengineered to help create transgenic animals. Researchers have taken out the deadly genetic material from the virus and have been able to use it to insert target genes into an embryo. This technique is more effective than earlier techniques and may be able to create transgenic birds and primates, which have been so far difficult to do.

BioDefinition

A **retrovirus** is a kind of virus that reproduces itself by transcribing its RNA genome into DNA, which can sometimes insert into the chromosomes of the infected cell. When the cell replicates, new copies of the virus gene are also made. Sometimes the retrovirus destroys the cells that it infects. This is done by the HIV virus that causes AIDS. In some other instances, the retrovirus causes the cell to become cancerous, as is the case with cancers that cause some leukemias.

- **Knockout technology** In some instances when you create a transgenic animal, you don't want to add a new gene to it—instead you want to "knock out" a gene. That is, you want to disable it so that it no longer functions. One way of creating a knockout animal—mice are the easiest to obtain—is to first genetically engineer an embryonic stem cell by disabling that gene. (An embryonic stem cell is a cell that has yet to differentiate into other cells, and can turn into any cell in the body.) Those bioengineered embryonic stem cells, often taken from mice with grey fur, are then inserted into developing mouse embryos, and the embryos are then surgically inserted into the womb of a mother with white fur. Some of the baby mice will contain the knockout gene and so will have patches of grey fur that come from the bioengineered cell. These mice are bred until a pure grey mouse with the gene knockout appears. This "knockout" animal will usually pass its knocked-out gene to its offspring.

- **Knockin technology** In knockin technology, a gene is not disabled, but instead its function is altered in some way. One way of creating a transgenic animal with knockin technology is to first genetically engineer an embryonic stem cell by altering the target gene. Then follow the same procedures for creating a knockout transgenic animal.

- **Cloning** Cloning techniques are an increasingly popular way to create transgenic animals. With this technique, the new gene is inserted into the DNA of an animal cell, not into an embryo. That animal cell, with the extra DNA in it, is then cloned by fusing it with an egg whose nucleus has been removed. The cloned animal will have the original DNA plus the new DNA inserted into the animal cell, and the embryo will start developing. When the gene is in the DNA of the developing embryo, it is passed along whenever cells divide so that all the cells of the animal have that new gene. In those animals that grow and thrive, the gene does its work—for example, producing the human protein. Not all cloned animals can pass their new genes to their offspring.

BioFact

As a safety precaution, some transgenic animals are made sterile so that they cannot reproduce and pass along their transgenic gene or genes to offspring. This is done so that if the animal gets loose from the laboratory, it cannot reproduce in the wild and pass along its genes to other animals.

How Transgenic Plants and Animals Can Be Used

Transgenic animals and plants have a wide variety of uses. We'll cover them in more detail in Chapters 20 and 23. In the meantime, here are some of the primary ways they are being used today and may be used in the future:

- **They can produce medicines at low cost.** Many kinds of pharmaceuticals cannot be produced in large quantities and so they are exceedingly expensive for the consumer. For example, a blood clotting factor called Factor IX that some hemophiliacs lack can cost between $100,000 and $200,00 a year. Transgenic plants and animals may be able to produce such medicines at low cost. For example, medicines could be produced in the milk of goats and then refined. There are many tests of this use of transgenic animals and plants, including one in which Factor IX is being produced in the milk of transgenic pigs. Vaccines against rabies, HIV, malaria, cholera and other diseases have been produced by transgenic tobacco, corn and potato plants.

- **They can produce industrial and other biochemicals at low cost.** Transgenic plants can produce industrial and other kinds of biochemicals potentially at low cost and with less environmental damage than when factories produce those chemicals.

BioFact

A variety of industrial biochemicals have already been created using transgenic plants. Trypsin, an enzyme used in quantities in the detergent and leather industries; and laccase, an enzyme used to manufacture detergents and fiberboard, have been produced by transgenic corn. The protein avidin, which is used to help purify other proteins, has been created by industrial crops since 1997.

BioWarning

Transgenic animals and plants can pose dangers as well as offering benefits—for example, they can potentially cause environmental damage, or be unhealthy to eat.

- **They can help cure world hunger.** Transgenic plants can be engineered to be more nutritious than nonbioengineered foods, have longer growing seasons, thrive in less-than-ideal conditions, resist disease, and be more nutritious than nonbioengineered foods.
- **They can help clean the environment.** Transgenic animals can be created that cause less pollution when they are farmed. Transgenic plants can be created that produce their own natural insecticides, cutting down on the need to use dangerous insecticides.
- **They can be used for organ transplants.** Many more people need transplants than can receive transplants because organs simply aren't available for everyone who wants one. Pigs are being bioengineered so that they can serve as organ donors to people.
- **They can aid in medical research.** Transgenic animals can also be used to test new medications and procedures. The best example of this is the transgenic "oncomouse" that develops cancer. Researchers can try different anticancer medicines and treatments and see if they are able to stop the mouse from developing cancer.

The Least You Need to Know

- Transgenic plants and animals are those in which the plant or animal has been given a gene from another plant or animal, or in which one or more of its genes has been disabled or altered in some way.
- Transgenic plants and animals are increasingly common, and transgenic plants in particular are in widespread use.

- In 2001, 68 percent of all soybeans grown in the United States were genetically modified.
- To create a transgenic plant, you transfer the target gene into plant cells, and then grow transgenic plants from those cells.
- There are a variety of ways to create transgenic animals, but in most of them, the target gene is inserted into an animal embryo in some way, and then the embryo develops into a full-grown animal with the new gene.

Chapter 20

The Debate over Bioengineering Animals

In This Chapter

- How transgenic animals can produce helpful pharmaceuticals
- How transgenic animals aid in medical research
- Other benefits of transgenic animals
- The potential health and environmental dangers of transgenic animals
- How the federal government regulates transgenic animals

We live in a world in which nature is a starting point, not an end point. Scientists now have the power to create animals never seen before on earth. They can take genes from one plant or animal and implant them in another animal, to give that new animal features or capabilities—for example, having ewes create medicine in their milk.

Scientists have given spider genes to goats, they've knocked out genes in animals so that those genes no longer work, and we're still at the infancy of the technology.

In this chapter, we'll look at the debate raging over the use of these bioengineered animals. We'll start off with a refresher course on how they're created and then discuss their pros and cons.

A Refresher Course on Bioengineering Animals

When we refer to bioengineered animals, we're talking about *transgenic animals*—animals that have received a gene from another species, or animals that have had a gene or more than one gene, knocked out, that is, disrupted so that it does not work properly.

> **BioDefinition**
>
> A **transgenic animal** is one that has received a gene from another plant or animal that it does not normally have. The animal may also have had one or more of its genes "knocked out," that is, altered so that it does not work properly.

Bioengineered animals can be used for many purposes. They may be able to produce useful proteins and drugs, or to be used for organ transplantation, as a food source, or in medical research.

The first step in bioengineering an animal is to isolate and determine what gene you want to give to it. If you want the animal to produce a particular human protein in its milk, you would first need to know what human gene is responsible for producing that protein. After you identified the gene, you would then need to isolate it outside the human body, so that you could implant it into the animal.

When you have the gene, you need to somehow get it into the animal. Genes are not implanted into adult animals. Instead, they are implanted into embryos. A variety of techniques, including microinjection techniques, are used to insert the gene.

Another method involving animal cloning starts by inserting the new gene into the nucleus of an animal cell, not into an embryo. The nucleus of that animal cell can then be fused with an egg whose nucleus has been removed. The egg has the original animal's DNA plus the new gene, and the embryo will start developing. All the cells of the embryo will have that new gene. When the animal is born, the gene can start doing its work—for example, producing the human protein.

In some cases, the transgenic animal will pass the new gene to its offspring, and in some cases it will not. Often traditional breeding techniques are used on several transgenic animals to breed one that always passes the new gene to its offspring.

BioFact

A new technique uses a harmless version of the AIDS virus to create transgenic animals and it may lead to a more efficient creation of transgenic animals. The AIDS virus and other retroviruses are able to copy genetic material directly into the chromosomes of an embryo. Researchers have been able to alter the virus by taking out its deadly payload, and instead, having it insert target genes into an embryo. Researchers say this technique is far more effective than earlier techniques and may also be able to create transgenic birds and primates, something that up until then was not possible.

How Bioengineering Animals Can Help Mankind

Bioengineered animals offer many benefits. They can potentially be used to:

- Produce drugs and pharmaceutical chemicals in large amounts at low cost.
- Produce nonmedical products and biochemicals at low cost.
- Provide organs for transplantation into human beings.
- Produce foods at lower costs than nonbioengineered animals.
- Help biomedical scientists uncover cures for diseases such as cancer.
- Make animal farming more environmentally friendly.

In the rest of this section, we'll look at their benefits.

Of Milk and Medication

It can be very difficult and expensive to produce many kinds of pharmaceuticals. Therefore, potentially life-saving therapies may not be available to all who want them, or may not be available at all. There are those who say that the use of bioengineered animals can produce these pharmaceuticals at low cost and may help save many lives.

The idea behind bioengineering animals to produce pharmaceuticals is fairly straightforward. First, the pharmaceutical to be manufactured is identified and then the gene or genes that helps create it are identified. Often the gene is modified to make it a better drug or to ensure that it is expressed in the correct cells of the animal.

When the gene is available, it is implanted into the embryo of an animal such as a goat or a sheep that produces milk. The animal produces the protein in its milk, and the protein is then purified from the milk and used as a pharmaceutical. The transgenic goats or sheep are often cloned so that entire herds of animals can produce useful proteins in this way.

BioFact

The first cloned mammal, Dolly the sheep, was cloned not for the fun of it. Rather, the company that cloned her, the Roslin Institute, wants to create transgenic animals that can produce medically useful human proteins. Dolly was just the first step in the process. Since the cloning of Dolly, the institute has been able to clone transgenic sheep. The first cloned transgenic sheep was called Polly.

For example, some hemophiliacs have genetic defects that do not allow them to produce a blood clotting protein called Factor IX. The protein can be produced to help them, but it is excessively expensive and costs between $100,000 and $200,000 per year per patient.

A group led by Dr. William Velander of the Virginia Polytechnic Institute and State University has created transgenic pigs that contain the human gene that can produce Factor IX. These pigs produce the protein in high concentrations in their milk. The work is still in its research stage, but Dr. Velander believes that ultimately only a few hundred of these pigs would be able to provide enough Factor IX for the entire world at a much lower cost than the current price of the protein.

This is just one small example. Many more abound—such as transgenic animals producing proteins that may be able to treat cystic fibrosis, heredity emphysema, and other diseases.

SpiderGoats, MilkSilk, and BioSteel

Transgenic animals will not be confined to producing pharmaceutical chemicals—they may produce other kinds of chemicals as well. In fact, they already have.

What do you get when you cross a goat with a spider?

If you're the biotech company Nexia Biotechnologies, you get BioSteel.

The company has produced a transgenic goat that contains genes from orb-weaving spiders. The goat produces a kind of spider silk in its milk and that silk can then be used to manufacture products. Spider silk is extremely strong, light, tough, and flexible. The company expects its silk to be used to create a variety of products, including fishing line, medical sutures, surgical meshes, and body armor. It has received a number of awards for its research, including awards from *Scientific American* and *Popular Science*.

Enter the Enviropig

Many people might not realize it, but animal farming can create major environmental problems. For example, there are many environmental problems caused by the large amount of manure generated by pig farms and chicken farms. Rivers and streams have been polluted with manure runoff.

BioFact

A good deal of research is being done into developing transgenic fish for a variety of purposes. Aquatic Systems, a division of Kent SeaFarms Corp., is using a $1.8 million grant from the Department of Commerce to develop striped bass that grow more quickly, require less feed, and are disease-resistant. Less feed and quicker growing time may mean less pollution from fish farming. As we'll see later in this chapter, though, transgenic fish, in particular a new kind of salmon, have become very controversial, with some claiming that the fish could be environmental hazards.

There are other environmental problems—for example, growing the amount of food required to feed animals such as pigs puts a major strain on the environment.

Pig farms are particular problems. Pigs require large amounts of phosphorous in order to live, but not all phosphorus in plants can be digested. Therefore digestible phosphorous is added to their feed. Pig manure contains high amounts of phosphorus and this ends up polluting lakes, rivers, and ultimately the oceans. The end result of phosphorous contamination is the killing of aquatic life.

Animals can be bioengineered so that they put less of a strain on the environment when farmed in large numbers. An example of this is a transgenic pig called the Enviropig. It can much more effectively digest phosphorous than nontransgenic pigs. It has been given a gene from a bacteria that produces the enzyme phytase, which helps digest certain forms of phosphorous. The Enviropig produces phytase in its saliva and it is able to digest up to twice the amount of phosphorous as normal pigs. Enviropigs' manure can have up to 64 percent less phosphorous than the manure of nontransgenic pigs.

Pig Farms and Organ Transplants

Consider these facts: Over 80,000 people are now waiting for life-saving organ transplants. Each day, some 17 people die because organs aren't available for transplantation. Every 13 minutes, a new person's name is added to the list of people waiting for an organ transplant.

Scientists believe that they have a way to alleviate the problem *and* save thousands of lives—create transgenic animals that can be used to grow organs that can be transplanted into humans. Research currently focuses on pigs for a number of reasons: they have organs approximately the same size as humans, and they are not so closely related to humans so that it is unlikely that a pig virus could jump the "species barrier" and infect humans.

A major problem with transplanting pig organs is that the human immune system attacks them because they are foreign to the human body. Therefore, transgenic pigs have been created that lack a gene that creates a sugar called gal. This sugar coats pig organs, and the human body attacks the organs because of the presence of gal. Without gal on the organs, the human immune system is less likely to attack and destroy the organs.

No organs from transgenic pigs have yet been transplanted into humans, but scientists believe that eventually they will and that pigs may supply many organs in the future. (For more information on about transgenic pigs and organ transplantation, see Chapter 10.)

Producing a Better Food Supply

Transgenic animals could conceivably produce a better food supply at less cost than the supply we currently have. Transgenic fish have already been created that grow more quickly than natural fish, and require less feed, which ultimately will mean lower cost. Animals can be created to have leaner meat, to be resistant to disease, and potentially to provide more nutrients.

Helping Medical Research

Currently, the widest use of transgenic animals is in medical research. They are used for a wide variety of purposes. They can help scientists understand the role that certain genes play in specific diseases. Researchers can introduce a gene into an animal, and then observe how the gene affects the animal. Or they can inactivate a gene in an animal and see what occurs. When you read or hear about scientists "discovering" that a gene may have been found that causes a particular disease or condition, a transgenic animal may have been used for the research.

BioFact

Here's one small example of how transgenic animals are used in medical research. Researchers at the biotech company GlaxoSmithKline bioengineered a transgenic mouse so that it produced an unnaturally large amount of the protein "uncoupling protein-3" (UCP-3), in the skeletal muscle in the mice. The mice ate more than mice who hadn't been engineered, yet were leaner, lighter, and had lower glucose and insulin levels. The researchers believe that regulation of the protein in humans may help treat obesity.

Transgenic animals can also be used to test new medications and procedures, to see whether they work, and whether they are toxic. A so-called transgenic "oncomouse" has been created that develops cancer. It was given a mutation in its MYC gene, which encodes proteins that regulate cell growth and cell differentiation. Because the mutated myc proteins can't function properly, the mouse develops cancer when some cells grow wildly out of control.

Because the mouse always develops cancer, researchers use it to test anticancer drugs and techniques, and see whether those techniques work. For more information about the oncomouse see Chapter 11.

BioFact

More than 95 percent of transgenic animals used in biomedical research are mice, according to the biotech company GlaxoSmithKline. Mice are ideal for this kind of manipulation and research because 80 percent of mouse genes are similar to those in humans; they have embryos that are easy to manipulate, so it is easy to create transgenic mice; they have a short reproduction cycle, so it is easy to grow them quickly; and they are easy to maintain in a laboratory.

Stopping Insect-Borne Disease

There are some who believe that transgenic animals may be able to help limit insect-borne diseases such as malaria. A transgenic version of the malaria-carrying mosquitoes could be created that would not allow it to carry the malaria parasite, and these might replace the disease-bearing mosquitoes.

Concerns About Bioengineering Animals

As you can see, there are many benefits offered by bioengineered animals. But there are those who worry that there are serious dangers that aren't being addressed, including environmental dangers and health hazards. In addition, there are those who have moral concerns about using animals in this way. In this part of the chapter, we'll look at the concerns that various people and groups have about bioengineered animals.

Environmental Hazards

Probably at the top of the list of potential dangers of bioengineered animals are environmental hazards. Specifically, many people worry what would happen to the environment and other plants and animals if bioengineered animals escaped into the wild.

It is not just environmental groups that worry about this kind of thing—many scientists have very serious concerns as well. For example, a special committee of the United States National Academy of Sciences issued a lengthy report about bioengineered animals that warned about the potential environmental dangers that they posed.

There are many potential environmental dangers, the report concluded. The transgenic animal, once in the wild, could start reproducing with its nontransgenic relatives, and so the natural population could eventually become transgenic. If the transgenic animal were fitter than the natural animal, the nontransgenic animals could eventually die out.

BioSource

To read the U.S. National Academy of Sciences report, titled "Animal Biotechnology: Science Based Concerns," go to www.nap.edu/catalog/10418.html. The entire 200-page report is available online, but you can only read a page at a time, so as a practical matter, you won't be able to print out the entire report unless you want to print it individual page by individual page. You can, however, order the book online from the site. An electronic version costs $21, while a printed version costs $34. You can also order single chapters in electronic format for $3.20.

A large population of transgenic animals could pose a danger to the environment. If they were bred to eat large amounts of food, or grow large quickly, they could lead to the destruction of other species as well. They could disturb the natural balance between predator and prey—if they were more effective predators, they could wipe out other species and upset the balance of nature.

The report issued a particular warning about transgenic salmon and other fish. Transgenic salmon have been created that can grow twice as fast as their natural counterparts, and work has been done on other transgenic-farmed fish, such as tilapia, carp, and trout. In some instances, the fish's genes are manipulated so that they produce high levels of growth hormone, and in other instances they have been bioengineered to be able to tolerate cold waters or to resist disease. Tilapia are also being bioengineered so that they produce human blood clotting factors as a way to produce pharmaceuticals.

The report concludes, "One case of immediate concern is the release of transgenic fish and shellfish." Although no transgenic fish have escaped into the wild, it notes, "Cultivated salmon have escaped into the wild from fish farms and these salmon already pose ecological and genetic risks to native salmon stock." Transgenic fish, the report warns, could pose even more of a danger because, for example, transgenic salmon have been created that grow four to six times faster than nontransgenic salmon, and they use up more oxygen and other resources than nontransgenic salmon.

BioSource

The Canadian novelist Margaret Atwood portrays a world in which biotech has run riot in her book *Oryx and Crake*. Food, people, viruses, and animals have all been bioengineered, leading to widespread ecological disaster and the collapse of civilization. People, for example, have been engineered so that they are docile, have a pleasant citrusy kind of smell, only need to eat grass, and no longer feel emotions such as lust or jealousy. Many transgenic animals roam the bleak landscape, including pigoons (a particularly lethal combination of pigs and baboons) and wolvogs (combination of wolves and dogs that look like friendly dogs, but have the killer group instincts of wolves).

The report warned that other transgenic animals, such as transgenic insects, cats, pigs, goats, mice, and rates, could pose environmental dangers as well.

Health Risks

There are potential health risks posed by transgenic animals as well, particularly if they are used in the food supply. A transgenic animal typically produces a protein or proteins not produced by the same nontransgenic animal and there is a possibility that people may be allergic to that protein. They would not normally expect that protein to be present in a pig, for example, and previously would not have had a problem when eating pork.

There is also the theoretical possibility that the transgenic animal could produce different proteins than scientists expect, and that these proteins could cause allergic reactions in people who eat the animal. The technique of microinjection, in which a new gene is injected into an embryo, could conceivably activate a gene in the animal that produces a toxin or allergen without the knowledge of the scientist.

There is also significant potential danger if a transgenic animal was created to produce pharmaceuticals, but then somehow made its way into the food supply. For example, let's say a transgenic animal were created in order to produce a specific human hormone in its milk. There is a chance that the transgenic animal would produce the hormone not only in its milk, but also in other tissues as well. If that animal mistakenly made its way into the food supply, there is the possibility that people could become ill or damaged from those hormones.

CAUTION

BioWarning

The possibility of eating transgenic animals is not a theoretical one—it may have already happened. At the University of Illinois, researchers had created transgenic pigs and then bred the transgenic pigs with one another. Between April 2001 and January 2003, university researchers sold 386 of the offspring to a livestock dealer, the pigs ended up being slaughtered, and entered the food supply. The researchers claimed that the pigs did not inherit the transgenic genes from their parents, and were not actually transgenic. The Food and Drug Administration, however, noted that the "FDA cannot verify this assertion because the researchers did not conduct sufficient evaluation or keep sufficient records to assess whether the offspring inherited the inserted genetic material."

Transgenic plants have become common in the food supply, but as of yet, there are no transgenic animals in our food supply. However, that is expected to change. Mike Wanner, former president of the biotech company Prolinia and chief operating officer of ViaGen which bought Prolinia, told the magazine *The Scientist* that he believes transgenic beef and pork will ultimately be a $2 billion business.

Another potential health hazard is that transgenic animals could pass along new viruses and other infections to humans. For example, a retrovirus that was used to deliver the gene into the embryo could combine with a latent virus in the animal's genome, or with a virus that later infects the animal—and this new recombined virus could then get into the food chain and infect humans. In the case of organ transplants from transgenic animals, viruses present in the animals' organs could infect humans.

Who Regulates Transgenic Animals?

Because of the potential dangers and benefits of transgenic animals, the federal government is paying attention—although depending on who you listen to, they're either paying too much attention or not enough attention.

The Food and Drug Administration (FDA) regulates transgenic animals. But some critics contend that the agency is hamstrung by federal law or by nature does not want to wield the regulatory authority it could. The Center for Science in the Public Interest says that the FDA has the proper tools in place for making sure that transgenic animals will be safe for human consumption if they ever reach market, but is ill-equipped to study and regulate any environmental damage that might be caused by transgenic animals.

> **BioSource**
>
> One of the best and most balanced overviews about the use of transgenic animals and their potential dangers is a report by the Pew Initiative on Food and Biotechnology titled "Biotech in the Barnyard: Implications of Genetically Engineered Animals." You can get it free at pewagbiotech.org/events/0924/proceedings1.pdf.

As this book went to press, a variety of bills were before Congress concerning the issue. Some would strip the FDA of authority, say critics, and others would give it more power to regulate transgenic animals. For more details see Chapter 15.

The Least You Need to Know

- Transgenic animals can be created that can produce medically useful proteins in their milk, and therefore may be able to create low-cost medications to cure diseases.
- Transgenic animals are helping medical researchers discover new treatments for cancer and other diseases and conditions.
- Some transgenic animals may help clean up the environment—for example, transgenic pigs that can better metabolize phosphorous may lead to cleaner rivers and lakes.
- Some people worry that transgenic animals can upset the natural balance of the environment if they get loose.
- There is a theoretical chance that eating transgenic animals can cause health problems, including allergic reactions for human consumers.
- The federal Food and Drug Administration regulates transgenic animals.

Chapter 21

Should We Bioengineer People?

In This Chapter

- A look at a future in which people are bioengineered
- Ways in which we can bioengineer people and create designer babies
- Building artificial chromosomes
- How preimplantation genetic diagnosis (PDG) can screen for genetic disease
- The pros and cons of PDG

Most discussions about genetics and cloning end up at the same place: Warning about the future as presented in Aldous Huxley's book *Brave New World*. He describes a dystopia in which humans are bioengineered to be one of five different castes: Alphas, Betas, Gammas, Deltas, or Epsilons. The Alphas are the ruling caste, and each caste below them is less physically and intellectually capable than the preceding one.

We are certainly a long way from *Brave New World*. But because of our increasing understanding of the functions of specific genes, and our ability to bioengineer life using those genes, there are those who worry that a world like one portrayed by Huxley may be in the making.

In this chapter, we'll look at the issues surrounding bioengineering people, including designer babies, and the ability to prescreen embryos for genetic disease.

The GenRich Versus the Naturals

It's several hundred years in the future. Human evolution has taken a strange turn—it's no longer left to mere nature and mankind has instead taken the reins.

There is no longer one kind of human being; instead there are two—the "Naturals" and the "Gene-Enriched," also called the "GenRich." The Naturals, which make up 90 percent of the population, are no different than you and me. They're handsome, homely, or in between. They're tall, average height, or short. Their intelligence, artistic skills, and athletic skills all fall within the ranges that we're used to.

In brief, the Naturals are much like you and I are today, and like our ancestors have been for millennia.

The elite 10 percent, the GenRich, on the other hand, are practically a different species. Like the imaginary children of Lake Wobegone, when it comes to the GenRich "all the children are above average." In fact, they're not merely above average; they're positively off the charts by today's standards. Each succeeding generation of the GenRich has had their genes carefully chosen and enhanced, and in recent generations, they have been enhanced by synthetic genes. All this costs a good deal of money, of course; what also separates the Naturals from the GenRich is their ability to pay for genetic engineering and the latest in, literally, designer genes.

The Naturals are trained from birth to do society's menial work; after all, with their skills so far beneath those of the GenRich, that's all they are good for. The GenRich elite are the scientists, artists, businesspeople, journalists, writers, and other members of the educated classes.

There is little contact between the two classes of people, and practically no interbreeding. In fact, the GenRich don't really breed in the traditional sense. Because their genes have become so specialized, there are genetic incompatibilities between them. Succeeding generations are engineered rather then bred.

Far-fetched? Perhaps. But that's a potential future envisioned by the Princeton microbiologist Lee M. Silver in his book *Remaking Eden: How Genetic Engineering and Cloning Will Transform the American Family*. Silver doesn't say this is inevitable, but warns that it's a possibility given the economic, social, and scientific trends of today.

Silver's vision may seem extreme, but it's a natural extension of plant and animal genetic engineering. In fact, as you'll see in the rest of this chapter, there are people today who believe that we've already begun to build "designer babies" and that it will only accelerate in the future.

BioSource

Remaking Eden: How Genetic Engineering and Cloning Will Transform the American Family is published by Avon Books and available in paperback for a list price of $14. For more information and a brief excerpt, go to www.harpercollins.com/catalog/book_xml.asp?isbn=0380792435.

Can We Build Designer Babies?

When we talk about *bioengineering* people, we often talk about building designer babies—being able to choose the specific attributes of our offspring, such as deciding to have a blonde-haired, blue-eyed baby who will have genes for high intelligence.

Doing that isn't possible today. But we have mapped the human genome and practically every day we gain a greater understanding of the purposes of specific genes and groups of genes. Armed with that knowledge, some believe, we will eventually be able to build designer babies.

BioDefinition

Bioengineering is sometimes used to refer to using engineering principles when treating the human body—for example, building artificial organs. In this book, however, we use it to refer to using genetic engineering as a way to accomplish a particular end, such as building a designer baby.

It's not yet clear exactly how you would go about building a designer baby, but here are a range of potential possibilities:

- **Screen an embryo for genetic disease.** You may not think of this as helping build a designer baby, but in essence, it is. The purpose of screening an embryo for genetic disease is so that you can discard the embryo if it has a life-threatening disease. In choosing to do this, you're choosing a child based on its genetic makeup. There's nothing wrong with this, of course, but it shows that to a certain extent we already choose designer babies. The technique that lets you

BioDefinition

Preimplantation genetic diagnosis (PDG) refers to a technique used to screen embryos for genetic disease. It is done on embryos that have been fertilized outside the womb before they are implanted into the uterus. If PDG finds out that the embryo has a genetic disease, it is not implanted into the womb, and another embryo, free of genetic disease, is instead used.

screen embryos for genetic disease is called *preimplantation genetic diagnosis* (PDG), and is being done today.

- **Select the gender of a baby.** There has already been some research on how you can make it more likely to have a girl than a boy or vice versa—by screening the sperm likely to produce males from those likely to produce females. A more foolproof way is to use PDG to determine the gender of an embryo and to then only use an embryo that matches the gender that you want. For example, when sperm fertilize eggs in fertilization clinics, the resulting embryos can be examined and only those that match the gender desired by the parents will be implanted. This capability is available today, although it is unclear how much, if at all, it is being used.
- **Select a baby with a genetic makeup similar to a sibling.** As we'll see later in this chapter, a number of parents have chosen to have children with a specific genetic makeup similar to that of one of their children with a genetic disease. Cells or tissue from the new child can be used to help cure the sick child—and the new child grows up to be healthy as well. This has already occurred several times.
- **Perform germ line therapy.** In germ line therapy, the chromosomes of an egg, sperm, or fertilized embryo would be altered to produce a baby with a specific genetic makeup. For example, a defective gene could be fixed in the embryo, and so the resulting child would not get the genetic disease. Preliminary work has already been done on a way of fixing the defective gene that causes Huntington's disease, a fatal degeneration of the nervous system. However, theoretically this type of therapy could be used to manipulate any gene or set of genes, and so genes could be manipulated to give children certain physical or mental attributes. This is the kind of therapy that most people refer to when they talk about "designer babies." Germ line therapy on humans is not yet possible, and not everyone is convinced that it will ever be possible.

BioFact

In the most primitive form, there has already been an attempt of sorts to create designer babies. Back in 1980, Robert K. Graham founded a "genius sperm bank" called the Repository for Germinal Choice in which women could pay to be impregnated with sperm from Nobel Prize winners and highly intelligent scientists. The only Nobel laureate to acknowledge making a sperm donation was William B. Shockley, the 1956 winner in physics. The sperm bank closed in 1999. No reports yet of Nobel-winning offspring.

Building an Artificial Chromosome

One way of building designer babies would be not merely to change existing genes in chromosomes, but to build artificial chromosomes and insert them into eggs or a developing embryo. That would mean that designer babies would have more chromosomes than other humans—and they could also pass that additional chromosome to their offspring.

Using artificial chromosomes could have a benefit over attempting to modify existing chromosomes. When you attempt to modify genes on existing chromosomes, there can be a number of different and dangerous unintended effects. For example, you may accidentally trigger a gene to begin functioning that you didn't want to function, or you may accidentally knock out the functioning of a gene. This can lead to serious medical problems, such as cancer—that is what has happened on a number of occasions in human gene therapy. (For more information about gene therapy, see Chapter 22.) Additionally, certain genes are very difficult to target.

BioFact

Researchers have discovered a gene that appears to be in part responsible for creating talented runners. The gene can make runners better at either endurance running or at sprinting, according to the version of the gene found in their chromosomes. The gene, alpha-actinin, produces a protein used by muscle fibers. One version of the gene, alpha-actinin-3 (ACTN3) is primarily found in sprinters, while another version, ACTN2, is usually found in long-distance runners. ACTN3 creates a protein used by "fast-twitch" muscle fibers, which are required more by sprinters. ACTN2 produces a protein used by "fast-twitch" muscle fibers, which are required more by endurance runners. The elite runners in both categories tended to have two versions of the gene, not one, which means that they would produce more of the protein.

With an artificial chromosome, however, you theoretically don't have those problems. Because the artificial chromosome is independent of the other chromosomes, the genes along it should work independently and should not disturb the functioning of existing genes. In theory, this extra chromosome could help cure genetic diseases. Genetic diseases are often caused when a mutation in a gene does not allow that gene to properly make proteins—and the lack of that protein's function causes the disease. You could insert a properly functioning gene on an artificial chromosome and insert that into an embryo, and in theory, it should be able to properly create proteins, curing the disease before it begins.

This is, of course, far into the future and no one is sure whether an artificial chromosome could work in humans or whether it could have dangerous side effects. Scientists have already created artificial chromosomes and inserted them into mice—and those chromosomes worked and were passed down from generation to generation.

The Canadian biotech company Chromos Molecular Systems created an artificial chromosome by starting with a natural chromosome and taking out most of its functional genes. They kept the parts needed for its replication, including the central X-shaped centromere and some so-called "junk" DNA between genes, which actually is not junk and serves important functions. The researchers then implanted genes into this stripped-down chromosome to create the new artificial chromosome. Once the chromosome was in the mice, it was passed down to offspring just as other chromosomes were.

BioFact

The researchers at Chromos Molecular Systems plan to use transgenic animals with artificial chromosomes so that they produce medically valuable proteins in their milk. One benefit that the artificial chromosome has over current ways of doing this is that with an artificial chromosome you can include many copies of the same gene. That means that the transgenic animal could in theory produce much more of the protein—if it has ten times the number of genes for creating the protein, for example, it could produce ten times the amount of the protein.

Mind Your PDGs

The closest technology we have today to building designer babies is preimplantation genetic diagnosis (PDG), in which embryos are screened for genetic disease. It is used only in special clinics that do *in vitro* fertilization. A number of eggs are fertilized outside the womb, and the embryos are then screened for genetic diseases. Only an embryo without genetic disease will be implanted.

It is an expensive proposition. For example, the well-known Reproductive Genetics Institute in Chicago charges $2,500 to screen an embryo, and that is in addition to the $7,500 for the *in vitro* fertilization procedure.

In PDG, one or two cells are removed from a three-day embryo, which frequently is made up of eight cells. The chromosomes from those cells are tested to see whether there are any abnormalities. There are a wide variety of abnormalities that can be tested for, including cystic fibrosis, Tay-Sachs, early-onset Alzheimer's, sickle-cell anemia, hemophilia, and muscular dystrophy. The cells are also tested to be sure that the appearance and number of chromosomes is normal.

CAUTION

BioWarning

PDG is not a foolproof technology. There is a chance that some of the cells in the developing embryo will have normal chromosomes, and others will have abnormal chromosomes—and it is not clear which of these types of cells will form the basis of the new fetus. At Wesley IVF Services in Australia, which performs PDG, a study uncovered that two patients for whom PDG had shown normal chromosomes eventually miscarried some time after the embryo had been implanted. Examination of tissue from the miscarriages showed that the embryos had, in fact, had an abnormal chromosome count.

Over 200 healthy babies have been born at the Reproductive Genetics Institute as a result of this testing.

As a general rule, PDG is only performed for specific reasons—if there is evidence that shows that the parents may pass down a genetic disease to their offspring. For example, a woman whose father, sister, and brother all developed early onset Alzheimer's, which develops when someone is in their forties, used PDG to screen for the disease. The screening was successful and an embryo was implanted in her that did not carry the form of the gene responsible for the disease.

Other Uses for PDG

PDG has been used for more than making sure that children are born free from genetic defects. It has also been used to select embryos that match a sick genetic offspring so that stem cells from the resulting baby's umbilical cord marrow can be used to treat the sibling.

There have been a fair number of cases in which this has happened, but the first instance was in 2000. Jack and Lisa Nash had a child named Molly who suffered from the rare genetic disease Fanconi anemia, which causes a variety of problems, including inadequate bone marrow production. Poor bone marrow production leads to leukemia and so Molly was likely to die within a few years.

If she were to survive, she would need a successful bone marrow transplant. If she were to receive a transplant from an unrelated donor, her odds of survival would be 42 percent. However, if she were to receive an umbilical cord blood transplant with stem cells from a sibling with compatible tissue, the odds would be more than doubled, to 85 percent.

CAUTION

BioWarning

PDG can also be used for purposes that some people believe are not necessarily ethical. For example, it has been used in some countries to screen embryos so the embryos implanted will be of a particular gender.

The Nashes used *in vitro* fertilization and PDG to select an embryo that was free of Fanconi anemia and that had tissue compatible with Molly's. The embryo was implanted, came to term, and a son named Adam was born. Molly received a transfusion of Adam's umbilical cord blood. It worked, and Molly appears to be healthy—and she has a healthy baby brother as well.

What the Naysayers Say

Despite the obvious benefits of PDG, there are critics who contend that the technique should not be used despite its obvious benefits. They wonder where the line will be drawn when using the technique, and whether it will be used to create a new generation of designer babies.

Jeffrey Kahn, director of the University of Minnesota's Center for Bioethics told the Associated Press, "Today it's early onset Alzheimer's. Tomorrow it could easily be intelligence, or a good piano player, or many other things we might be able to identify the genetic factors for."

Although Kahn doesn't call for banning the procedure, he believes that there needs to be a policy discussion about how it can be used, and regulation or oversight over the procedure.

The Cons of Designer Babies

There are many others who agree with Kahn—and some call for an outright ban on certain uses of PDG. They also warn against the potential effects of being able to create designer babies. Here are some of their main concerns:

- **It will be used as a form of eugenics.** There was a movement in the twentieth century that in essence wanted to make sure that people and races with "bad" genes or attributes did not pass those attributes to succeeding generations, and that those with "good" attributes did pass them along. One end result of this was Nazism, the extermination of most of the European Jews and an attempt to create a master race. In this view, PDG and similar techniques—especially being able to bioengineer human beings—is another form of eugenics.

- **It makes us "play God."** There are certain lines that should not be crossed by human beings, and one of those is manipulating genes, and choosing who shall be born and who shall not.

- **It can be dangerous for the developing child.** PDG is not believed to be dangerous to the subsequent developing children or adult. But actual genetic manipulation of embryos could have dangerous, unintended consequences that we do not know about and that should not be attempted.

BioFact

The argument of whether people should "play God" was turned on its ear by Nobel winner James Watson, co-discoverer of the structure of the DNA molecule. Testifying before the British Parliamentary and Scientific Committee he asked, "If scientists don't play God, who will?"

- **It can create several classes of people.** As Lee M. Silver warned, only those with money will be able to afford genetic enhancements. That could ultimately lead to two classes of people—the genetic haves and the genetic have-nots.

- **It treats children as commodities.** If parents can choose a child based on gender, hair color, eye color, and similar attributes, then children become little more than commodities.

- **It results in the discarding of embryos or abortions.** PDG means that certain embryos will be discarded, and genetic testing in the womb could lead to parents deciding to abort embryos with genetic defects. Those opposed to abortion oppose this.

- **It can do harm to the genetic makeup of society as a whole.** There are those who say that if we bioengineer people, it will lead to a less varied society made up only of those who have certain physical and mental attributes.

The Pros of Designer Babies

When people speak in favor of designer babies, they are not, as a general rule, speaking in favor of being able to genetically enhance humans or give them synthetic chromosomes. Rather, they are in favor of techniques such as PDG that can help stop children being born with genetic diseases.

> **BioSource**
>
> The American Association for the Advancement of Science has prepared a report on the scientific, religious, ethical, and policy issues around human inheritable genetic modification—in other words, bioengineering people by changing their genetic makeup, and having that genetic makeup be passed down to subsequent generations. Get a copy of the report for free at www.aaas.org/spp/sfrl/projects/germline/report.pdf.

Their arguments are straightforward: It is unfair to the yet-to-be-born to allow them to be born with debilitating and deadly diseases. Just as doctors attempt to cure people after they're born, the thinking goes, they should also attempt to make sure that people are not born with deadly diseases in the first place.

They also point out the benefits of being able to choose a child to be born who is a tissue match for a sick sibling—and whose umbilical cord blood can be used to save the life of that sibling.

Additionally, they note that in many cases miscarriages occur because of genetic defects in the embryo, and PDG in that sense is only doing what nature already does—selecting the embryos that are free of disease.

Some also say that parents have reproductive rights to choose whatever baby they want—and that even extends to choosing whether to have a boy or a girl.

Finally, there are those who believe that science should be as unfettered as possible and placing limits on the development of reproductive technology is a bad idea.

The Least You Need to Know

- Designer babies could be created in a number of different ways, including via germline therapy, which alters the genes of a sperm, egg, or embryo.
- Human germline therapy is not yet possible, but artificial chromosomes have been created and successfully implanted in mice.

- In preimplantation genetic diagnosis (PDG), an embryo fertilized outside the body is tested for genetic defects, and only those embryos that are free of genetic defects are placed in a womb so that they can grow into babies.
- Some people oppose PDG, claiming that it is the first step toward the creation of "designer babies."
- Those who oppose designer babies and PDG say that it can be used a form of eugenics, in which only those with "good" genes are allowed to live.
- Those who favor PDG note that it can protect against deadly and debilitating genetic diseases.

Chapter 22

The Promise and Perils of Gene Therapy

In This Chapter

- How genetic disease is caused
- The ways in which gene therapy can help cure disease
- How gene therapy works
- How gene therapy has been used to combat Bubble Boy Syndrome
- Potential dangers with gene therapy
- Ethical issues and gene therapy

Increasingly, scientists have been uncovering the genetic roots of many diseases. A wide range of diseases, from Bubble Boy Syndrome to cystic fibrosis, hemophilia and many others, are caused by mutated genes. And we're only now understanding how many other diseases are caused by genetic predispositions.

In the past 10 to 15 years, scientists have been developing a promising way to combat some of these diseases—using gene therapy. In gene therapy, a healthy gene is inserted into the body, and does the work that a mutated gene cannot. In this chapter, we'll look at what gene therapy is, how it works, and the benefits and dangers associated with it.

What Is Gene Therapy?

Before you can understand gene therapy, you need to first review some gene basics that you learned earlier in the book, and understand how genetic diseases are caused.

Bad Genes: The Cause of Genetic Disease

Genetic diseases can occur when there are mutations in someone's genes that make them malfunction. A single gene, remember, contains instructions for manufacturing a protein or family of proteins (often enzymes). Proteins build bodies and keep them functioning properly. So a genetic mutation can cause proteins to be manufactured improperly, or not be manufactured at all. Sometimes mutations can produce proteins that are dangerous. When any of those things happen, disease can occur.

For example, the disease hemophilia is caused by a mutation in a single gene. In the most common form of hemophilia, hemophilia A, the FVIII gene found on the X chromosome, is mutated so that it cannot properly produce a protein called Factor VIII that helps the blood clot. Because the protein isn't manufactured correctly, the person's blood cannot clot properly, and so when he is cut, the bleeding does not properly stop.

BioFact

Because the gene that causes hemophilia is found on the X chromosome, it is rare for females to get the disease. Remember, an XX chromosome pair results in a female, and an XY pair results in a male. That means a male has only one set of chromosomes that can cause the disease—the single X chromosome. If that single X chromosome contains the mutated gene, he will get hemophilia. However, because a female has two X chromosomes, she has two copies of the chromosome, and so two copies of the gene. Because the mutated gene is rare, the odds are one of those genes will work properly to produce clotting factor and so will be dominant over the mutated one. So females rarely get hemophilia.

Commonly, genetic diseases are inherited; the mutated gene is passed from one or both parents to their offspring. Hemophilia, for example, is commonly inherited. However, in some cases the disease is not inherited—the gene mutates when the sperm or egg cell is created. In some rare instances, hemophilia is caused this way rather than being inherited. But in either case the mutant gene will be passed on to the afflicted person's children.

Keep in mind that both environmental and genetic factors often cause disease. That means that someone can be born with a set of genes that predisposes him to a particular disease, but depending on the way the person lives and his environment, he may or may not develop the disease, and the disease may be mild or serious.

In fact, there are four types of genetic disorders:

- **Single gene (also called monogenic or Mendelian).** In this type of genetic disease, the mutation of a single gene causes the disorder. Hemophilia is this type, as are cystic fibrosis, sickle cell anemia, Huntington's disease and others. Because a single mutated gene causes these diseases, single-gene diseases have been the focus of gene therapy and gene therapy research.
- **Multifactoral (also called polygenic or complex).** This type of genetic disease is caused by a combination of environmental factors and genetic mutations. For example, genes that influence whether someone will get breast cancer have been found on chromosomes 6, 11, 13, 14, 15, 17, and 22. Many common diseases and conditions are multifactoral, including high blood pressure, heart disease, diabetes, and obesity.
- **Chromosomal.** In some instances, genetic disease is caused by abnormalities that are visible simply by looking at chromosomes with a microscope, for example if a chromosome is shorter or longer and therefore contains an abnormal complement of genes. The most common type of chromosomal genetic disease is Down Syndrome, caused when a person has three copies of chromosome 21 instead of the normal two.
- **Mitochondrial.** Mitochondria, as you learned in Chapter 3, are the powerhouses of the cell, and are located outside the nucleus. They contain their own DNA, however, and mutations in mitochondrial DNA can cause diseases. Pearson Syndrome, in which there are bone marrow and pancreas dysfunctions, is a kind of mitochondrial genetic disease.

BioSource

If you have a genetic disorder, know of someone with a genetic disorder, or want to find more information about genetic disorders, a good source of information is the Genetic Health Alliance at www.geneticalliance.org. It includes information on the disorders, as well as where and how to get help.

How Gene Therapy Might Help

Because genetic diseases are caused by faulty genes, scientists hope that fixing those genes might cure or alleviate the symptoms of those diseases. Unfortunately, there is no way to simply correct DNA by eliminating the mutation. Instead, one tries to add back an extra copy of the gene, one that functions normally. Rather than try to cure the disease with drugs, the body itself would provide the cure—its genetic manufacturing plant would swing into action and fabricate the proper proteins. So, for example, in the case of hemophilia, a normal FVIII gene would be incorporated into the cell nucleus, and it would manufacture Factor VIII, and so blood would clot properly. The hemophilia would be cured.

Gene therapy is aimed at the mature cells of the body—what are called somatic cells. It isn't being tested, for now at least, on changing the genes of egg cells and sperm cells. In other words, the current thinking is not to check egg cells, sperm cells or embryos for genetic mutations, and somehow fix them by adding a normal gene. This is not only complex and difficult but raises ethical questions because it could change genes in a way that will be inherited by future generations. Rather, gene therapy is aimed at fixing the problems caused by faulty genes in an already developed human being.

In gene therapy, a copy of a normal gene is inserted into the nuclei of cells in the hopes that a normal gene will take over the functions of the mutated gene—in other words, that the normal gene will produce the proper protein or family of proteins. With the body producing those proteins on its own, the person should be cured. As a practical matter, the healthy gene can't possibly be inserted into the countless number of cells that make up the human body—and in fact, there's no need to have the gene in every cell. Remember, only certain cells produce certain proteins. So the healthy gene only needs to be inserted into those cells responsible for producing the proper protein. In the case of hemophilia, the healthy gene would only need to be inserted into those cells responsible for producing Factor VIII, which helps clot blood.

All this sounds simple in theory. But as we'll see later in this chapter in the section "How Gene Therapy Works," the practice of gene therapy is a lot harder than theory. There are a great many hurdles that have to be leaped before gene therapy is a safe means of curing disease. It is still an experimental therapy, and to a certain extent a controversial one as we'll see later in this chapter.

Helping to Fight Cancer?

Gene therapy could be used in other ways as well, not just for fighting genetic disease—it could potentially be used, for example, to fight cancer. Although there are many types of cancer, some cancers can be caused when a "suicide gene" in a cell no longer functions. Often, if something goes wrong with the DNA in a cell, a "suicide gene" springs into action and kills the cell. This not only makes sure that only healthy cells in our body survive, but it also makes sure that cells can't reproduce out of control. But if the suicide gene malfunctions, cells with abnormal DNA won't die, and can multiply out of control. In that case, the cells can keep growing, and become a cancer. Gene therapy could potentially be used to turn that suicide gene back on in the cancerous cells, and the cells would not only stop multiplying, but the existing cells would die. At least that's the theory.

CAUTION

BioWarning

Although some scientists believe that gene therapy can help cure cancers, the therapy has already been shown to have the potential to cause cancer. It can cause cancer because when the healthy gene is inserted into cells, the gene may land on a location on chromosomes that leads it to "knock out" a different healthy gene, and cause that gene not to function properly, leading to cancer. We'll look at this in more detail later in this chapter.

How Gene Therapy Works

There has been a great deal of experimentation with gene therapy, and so many different methods developed for it. However, the general way it is done is similar in most methods.

First, the mutated gene that causes a disease must be identified, so that scientists can know which gene to target. Next, scientists have to isolate a healthy copy of that gene—and that healthy copy is what they'll inject into the person with the disease.

After they have copies of the healthy gene to be injected, they have to get the gene into the body—and they have to target the gene so that it is used by the specific cells whose job it is to create the protein or family of proteins. Injecting DNA into a cell isn't a simple matter of putting a needle into an arm. If you just injected DNA into the bloodstream, it wouldn't be able to be used. The DNA wouldn't be get into cells, and it certainly wouldn't be targeted at specific cells.

So what to do? The DNA is packaged into what's called a vector. The vector's job is to deliver the healthy DNA into the nuclei of the specific cells of the patient—those cells that will manufacture the protein that the body needs to become healthy. This is a tall order, as you might imagine. If we can't even figure out a way for the pizza delivery guy to send pepperoni pizza to the right address, how are we going to deliver healthy DNA into the nuclei of only specific cells?

As it turns out, there is an ideal vector. And that ideal vector can also be one of the human body's worst enemies—viruses. They're the ideal vector because of the way they work. In some ways, *viruses* are more machines than they are living beings. In essence, a virus is just a protein sheath wrapped around a bunch of DNA or sometimes RNA. When a DNA virus invades the body, it latches onto a cell and then injects its DNA into that cell. The virus's DNA hijacks the machinery of the cell, and uses it to produce more of itself—it makes many copies of the virus. In the worst cases, so many copies are made that the cell in essence explodes, all the copies of the virus get loose in the body, and each of them in turn infects a new cell. Then the cycle starts all over again, with each of those cells producing many viruses, and each of those viruses in turn invading new cells.

BioDefinition

There are many different kinds of **viruses**, categorized by the way in which they work. You may have heard the term retrovirus, because human immunodeficiency virus (HIV), the virus that causes AIDS, is a retrovirus. Instead of carrying DNA, a retrovirus contains RNA, along with an enzyme that uses the RNA to synthesize a DNA molecule. The retrovirus injects the RNA and the enzyme into a cell, and once inside the cell, the enzyme uses the RNA to synthesize DNA. That DNA in turn hijacks the machinery of the cell and uses it to create many new copies of the retrovirus, which then go on to infect new cells.

There's a saying that when life hands you a lemon, make lemonade. So when life handed scientists this exceedingly efficient disease-maker, scientists figured they could put that efficiency to their own use. Because viruses are so good at injecting DNA into cells, they reasoned, why not use viruses for gene therapy? Why not hijack the virus, and force it to inject healthy genes into cells that have unhealthy genes?

That's exactly what they did, and that is the most common way that gene therapy works today. The virus is bioengineered so that it can't reproduce itself. Then scientists place the DNA that they want to deliver to the human inside the virus. So when the virus comes across a human cell, it does what a virus normally does—it attaches to it and injects DNA into it. But that DNA is not the normal viral DNA. Instead, it contains the healthy human gene. The gene enters the cell's nucleus, and there it functions like a normal healthy gene, and manufactures the protein that the body needs to become healthy.

The bioengineered viruses are not put directly into the body. Instead, the viruses are usually mixed with the patient's target cells outside the body. Those target cells take up the healthy genes from the viruses. After they take up the healthy genes, they are placed into the appropriate tissue or organ, and then they manufacture the needed proteins. This means that only certain cells in the body will have the healthy genes—the cells that do the actual manufacturing of the proteins.

Earlier in the chapter, we reviewed the four types of genetic disease. Currently, gene therapy is only being used to combat one of those types—single gene (also called monogenic or Mendelian). In this type of genetic disease, the mutation of a single gene causes the disorder, and it's an ideal candidate for gene therapy because only a single gene needs to be inserted.

Other Types of Gene Therapy

Using a virus is currently the most popularly accepted way to deliver healthy genes into the nuclei of cells. But there are other ways that are being investigated as well. In a kind of brute-force approach, large amounts of DNA are directly injected into the tissue that could produce the protein. The DNA is taken up by the tissue, and the tissue then produces the protein. There are a number of drawbacks to this—only certain tissue can be treated this way, and very large quantities of DNA are required because only a tiny fraction of the DNA survives the treatment.

BioFact

One of the more amazing types of gene therapy that scientists are considering is building an artificial chromosome and delivering it into cells. It would be a 47th chromosome (the human body normally has 46), and would exist alongside all the other chromosomes. It would function like any other chromosome and so would have healthy genes that would produce the needed proteins. This would be useful for instances in which more than one mutated gene causes a disease.

In another type of gene therapy, liposomes, fatty particles, could be used to deliver the gene rather than a virus delivering the gene. The liposomes essentially dissolve the DNA and can deliver it to cells. So, in this type of therapy, the gene is transferred to the cells inside the patient's body, unlike when viruses are used, in which case the gene is transferred to cells when the cells are outside the patient's body.

Gene Therapy and the Bubble Boy Syndrome

Although gene therapy sounds simple, as you might imagine there are many complications involved. Gene therapy trials have been going on for more than a decade, and there are still problems with them and they remain controversial. So the best way to get a better understanding of gene therapy is to look at it in real life.

BioDefinition

The official, scientific name for the **Bubble Boy Syndrome** is X-SCID, severe combined immunodeficiency.

The most comprehensive, and the most controversial gene therapy trials involve the so-called *Bubble Boy Syndrome*, a rare disorder in which someone's immune system is so damaged by genetic disease that it can't fight off infections, and so the only way to survive is to live in as an antiseptic environment as possible. Typically, those born with the disease die by the time they are two years old.

The disease is caused by a mutation of the IL2RG gene on the X chromosome. That gene creates an important immune system protein called interleukin-2, and so people who have the syndrome can't manufacture it. Bubble Boy Syndrome is an ideal candidate for gene therapy, because it is caused by a single gene.

In the latest trials, begun in Paris in 2000, a copy of healthy IL2RG genes were packaged into bioengineered viruses—the vector that would insert the healthy gene into the cells of the children with the disease. The cells that produce interleukin-2 are located in the bone marrow, and so bone marrow cells are taken out of the body of the children with the disease, and mixed with the viruses that have the healthy gene.

BioFact

Girls rarely get Bubble Boy Syndrome, because it is caused by a gene found only on the X chromosome. The mutation is recessive, which means that if there is one healthy gene present, interleukin-2 can be produced. Because boys have only one X chromosome (they have one X and one Y chromosome), if the gene is mutated on the chromosome, they will get the disease. Girls, however, have two X chromosomes, and so two copies of the gene. The great odds are that one of those two copies of the gene will be healthy, which is why they are not likely to get the disease.

The healthy DNA is injected into the bone marrow cells. These cells are then put back into the children in a special germ-free isolation room. The cells with the healthy DNA, back in the children's body, start producing interleukin-2, and so they are able to fight off disease.

Initially, the results were successful. Children treated this way developed healthier immune systems and were able to fight off diseases. The trials have been carried out in France and in England, and the mother of one of the children in London said that the results were "nothing short of a miracle," according to the *New Scientist* magazine.

However, problems developed—several of the children developed leukemia. Because of this, the trials were halted in France, although they were still carried on in Great

Britain where it was argued that the children would die quickly without gene therapy. We'll look in more detail at why the children developed leukemia in the section "The Trials to Date: Is Gene Therapy Too Dangerous?" First, though, we'll look at some other gene therapy trials and research.

BioFact

The first gene therapy trial approved for treatment was for adenosine deaminase (ADA) deficiency. This rare genetic disease is caused when a mutated gene cannot properly produce the ADA enzyme. The trials began in 1990, and the children involved had to receive gene therapy several times over a number of years. Tests in more recent years have shown that the re-engineered cells are still producing the ADA enzyme. However, the children also take small doses of a drug called PEG-ADA to keep their disease under control, because the bioengineered cells cannot produce enough ADA by themselves to keep the children healthy.

More Ways That Gene Therapy Is Used

Gene therapy for the Bubble Boy Syndrome is only one of many trials and areas of research involving gene therapy. There is a flood of research and news involving gene therapy; here are some of the more recent developments:

- A research team at UCLA has been able to get genes into the brain by using liposomes, which can be engulfed by the cell along with the DNA carried by them. Viral vectors can't cross the "blood-brain" barrier, and so some believe that liposomes may be able to be used to treat brain disorders, such as Parkinson's disease.
- Scientists at Baylor College of Medicine in Houston used gene therapy to cure mice of diabetes, so that the mice produce insulin. The scientists believe that eventually a similar technique may be able to be used on humans.
- "Gene silencing," also called RNA interference, may be able to be used to treat Huntington's disease, a fatal brain disorder. In gene silencing, a kind of RNA called siRNA (short, interfering RNA) is used to interfere with the production of the abnormal protein that causes the disease.
- Gene therapy may ultimately hold a cure for cystic fibrosis, a lung disease in which a mutated gene causes mucus to be overproduced, clogging the lungs and leading to chronic infections. A healthy form of the gene would be delivered via an inhaler. Trials have been done, and have not proved to have long-lasting effects, but new techniques may make the therapy more effective.

All this, of course, is the good news. Unfortunately, there's been bad news as well, including deaths that have led to the cancellation of gene therapy trials. In the next section, we'll look at the problems with gene therapy, and why they occur.

Is Gene Therapy Too Dangerous?

Gene therapy started out with great fanfare and promise more than a decade ago, and there were great hopes that it might soon cure a wide range of genetic disease. But while many scientists still believe that it will eventually cure them, it has proved to be a dangerous therapy as well—so dangerous, in fact, that some gene therapy trials, as of this writing, have been halted.

> CAUTION
>
> **BioWarning**
>
> The first patient to die from genetic therapy was 18-year-old Jesse Gelsinger, who died in 1999 when undergoing gene therapy for ornithine transcarbamylase (OTC) disorder, in which the liver cannot function properly to clear ammonia from the bloodstream.

The problem is not only that the therapy can be dangerous, but there are also a good many medical and technical hurdles it has to overcome before it can become useful as well. These are among the most important ones:

- **Gene therapy is generally short-lived.** The genes inserted into the body often go to cells that eventually die and may be replaced by cells that do not have the new gene, so the benefits go away. So it's likely that people will have to undergo many rounds of the therapy in order for it to work.
- **It is difficult to target the right cells and tissues.** It's not always possible to get the healthy gene into the specific cells and tissues that will in turn produce the proper protein.
- **Gene delivery systems don't always work.** Scientists are still experimenting to find out which vectors are effective. There are a variety of different viruses and other methods that are being tested, but it's still not clear which will work best.
- **The body may attack the vector.** Our immune system is designed to attack any foreign invaders, and a viral vector is a foreign invader. So the body may use its built-in defenses to try and kill virus carrying the gene.
- **Many genetic diseases are caused by disorders of multiple genes.** The most common genetically related disorders, such as high blood pressure, heart disease and others, are caused by problems with multiple genes. It may be impossible to be able to treat these multigene disorders with gene therapy because of the complexities involved.

Gene Therapy and Cancer

Even more problematic for gene therapy than the issues just covered is the link between gene therapy and cancer. Trials for the Bubble Boy Syndrome were halted in France when several children undergoing the gene therapy developed leukemia.

To understand how gene therapy can cause cancer, we'll have to look a little more closely at how gene therapy works. When the viral vector injects the healthy gene into a person's cells, the gene doesn't latch itself onto the chromosome at the specific spot where the mutated gene is located. Instead, it stays in the nucleus or at a random spot on a chromosome, and then uses the nucleus's normal protein-making machinery to manufacture the needed protein.

BioSource

For excellent background information about how gene therapy works, go to www.ornl.gov/TechResources/Human_Genome/medicine/genetherapy.html, run by the U.S. Department of Energy Office of Science.

However, scientists do not yet have a way of controlling where the gene actually lands—and that's where the problems come in. The gene can essentially knock out the functioning of another gene if it lands on it or next to it on the chromosome, or could turn on the functioning of a gene if it lands on it or next to it on the chromosome. And turning certain genes on or off can cause cancer. For example, if the inserted gene turns off one of a cell's "suicide genes" it can cause cancer. These genes kill the cells if they threaten to reproduce in an out-of-control fashion. If that suicide gene is turned off, the cell can divide wildly and lead to cancer.

In the case of gene therapy for the Bubble Boy Syndrome, apparently in several instances the healthy gene was inserted next to a gene called Lmo2 that when triggered can cause cancer. Apparently the insertion turned on Lmo2. So two of the 11 boys in the trial developed leukemia, and it is possible that others might develop the disease as well. The trials in France were halted and so were similar ones in the United States.

Further research has shown that genes landing next to other genes, rather than somewhere else on the chromosome, are not accidental. In fact, apparently genes carried by retrovirus vectors are to a certain extent predisposed to landing next to other genes. A study published in the journal *Science* studied two viruses and found that one landed on genes 34 percent of the time, and another 58 percent of the time. If the viruses had merely fallen randomly, they would have landed on genes 22 percent of the time, the study said.

It's unlikely that gene therapy will be permanently banned because of this or similar problems, but it does mean that it will go more slowly and cautiously than previously.

The Ethics of Gene Therapy

The use of gene therapy raises a number of ethical issues. Clearly, curing a deadly disease is good, but things are not always that clear-cut when it comes to gene therapy. Here are some of the major ethical issues relating to the therapy:

- **How much of a risk does the therapy pose?** Can the cure be more dangerous than the disease? And even if there are risks, shouldn't they be taken, if a disease can be cured? For example, in the Bubble Boy tests, if gene therapy wasn't used, the children most likely would have died, or had a very poor quality of life. Is it worth the risk of developing leukemia—a treatable disease—to cure a deadlier disease?
- **Can the patient give informed consent?** In the case of Jesse Gelsinger, who died from undergoing gene therapy, his parent filed suit, saying he hadn't been adequately warned of the dangers inherent in the trials. In highly experimental procedures like gene therapy, is there even such a thing as informed consent if the dangers aren't fully known?
- **What is the line between curing disease and making genetic "improvements"?** No one disputes that if we can cure genetic diseases such as Bubble Boy Syndrome and diabetes, we should. But what about making genetic "improvements"? If we can genetically modify a developing embryo to be "normal" that would otherwise result in Down Syndrome, should we do it? If so, should we genetically modify an embryo that is genetically predisposed for a low IQ and raise its IQ if we can? How about parents who want blue-eyed children, or children who will be athletically talented—if we had the power to do that, should we do it? We covered these and similar issues in Chapter 21.

The Least You Need to Know

- Genetic diseases are caused when someone's genes malfunction so that they cannot produce a protein or family of proteins, or so that they produce a protein that is dangerous.
- In gene therapy, a copy of a normal gene is inserted into the nuclei of cells so that the normal gene will take over the functions of the mutated gene and produce the proper protein or family of proteins.
- To insert the healthy gene into a cell, a vector such as a virus is often used. The vector injects the gene into the cell, where it is taken up in the cell's nucleus and can function, just like any other gene.

- Gene therapy can potentially cause disease if the inserted gene lands near a cancer-causing gene and turns that second gene on or off.
- Trials of gene therapy for Bubble Boy Syndrome were halted because two children developed leukemia.

Chapter 23

BioFoods: Frankenfoods or a Cure for Hunger?

In This Chapter

- A look at BioFoods and genetically engineered crops
- How to bioengineer foods and crops
- Where to find bioengineered crops
- How bioengineered crops may turn farms into manufacturing facilities
- The pros and cons of BioFoods

The battle lines cannot be any clearer. On one side, biotech firms and many scientists claim that genetically engineered crops and foods may help cure world hunger and perhaps help clean up the environment. On the other side, environmentalists and scientists warn that we are creating foods that can help destroy the environment and may well be harmful to human health.

In the middle, there's us. In this chapter, we'll take a look at the controversies surrounding genetically modified foods and crops. We'll also look at how the crops are created, and at how widespread they have become—and you'll be surprised at how much they have become a part of your daily life.

Take One Gene from Column A ...

Since mankind has been farming, it has been looking for ways to improve its crops. Virtually none of the fruits, grains, and vegetables you eat was created by nature; over millennia we have bred and crossbred them so that they have specific characteristics we want and are missing specific characteristics we don't want.

For example, what we call corn was originally a wild grass native to Central America, called teosinte, and did not have the large cob and kernels that we identify with corn. Native Americans collected and cultivated the grass, breeding it until it became maize, the precursor to corn. All other food crops have been similarly bred through the centuries.

BioFact

Teosinte looks very different from the corn we know today. It has no cob, and its "kernels" are tiny green grass seeds. But according to the geneticist George Beadle, only five genes are responsible for most of the major differences between the two plants. For example, the teosinte's tga gene is responsible for forming a hard casing around each kernel. In corn, however, after generations of breeding, the tga gene was eventually modified so that the casing doesn't enclose the kernel. The Teosinte's tb1 gene causes the plant to have many stalks per plant, while the corresponding gene in corn limits the number of stalks. With too many stalks, the ears of corn cannot each become large.

Developing crops this way is called selective breeding. Farmers select plants that have the qualities that they want future generations of plants to have, breed those plants with one another, and over many generations of plants, are able to produce plants with the given characteristics. In doing this they do not work that much differently than Gregor Mendel, the father of genetics. (For more information about Mendel and genetics see Chapter 2.)

The ultimate goal of bioengineering plants is the same as the goal of selective breeders—to create better plants. However, the way they go about doing it is very different. Rather than take a hit-and-miss breeding approach, in bioengineering a gene is first identified that will produce a specific protein or family of proteins that will give the plant a particular characteristic. That gene is then isolated and spliced into the plant. The plant then exhibits the desired characteristic.

There's another very big difference between traditional breeding and bioengineering—the gene spliced into the plant need not be originally from a variant of the plant. In fact, it rarely is. For example, a fish gene can be implanted in a tomato—and, in fact, that has already happened in one of the earliest examples of plant engineering. Researchers discovered a gene in the arctic flounder that produced a protein that acts a kind of natural antifreeze, allowing the fish to survive in the cold Arctic waters. They implanted that gene into a tomato plant in the hopes that they could produce a tomato that could be subject to freezing temperatures and would still retain its texture after being defrosted. The experiment only partially worked. The tomato did end up producing the natural antifreeze. However, it had no effect on allowing the tomato to retain its texture after being frozen.

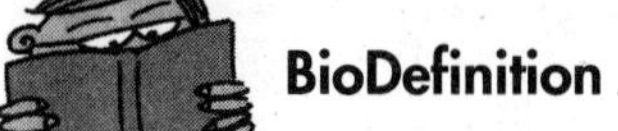

BioDefinition

A **transgenic plant** is one in which a plant has received a gene from a different species—from either an animal or another plant. If genes have been removed from a plant, it is also referred to as transgenic. Transgenic plants and crops are frequently also referred to as GM crops, shorthand for genetically modified.

How to Create a Transgenic Plant

In Chapter 19 we covered, in detail, how *transgenic plants* are created. But here's a brief refresher course.

Let's say you want to create a transgenic plant that is resistant to herbicides. First you identify the gene that produces a protein that will make the plant resistant. Then you implant that gene into the DNA of cells from a plant using a variety of different methods. The gene for herbicide resistance becomes part of the cells' genetic material. The cells are then grown into whole plants, which will be resistant to herbicides applied to fields to kill weeds.

The Most Popular Transgenic Plants

Transgenic plants can be created for a variety of different purposes, for example, to create plants that have added nutritional benefits, that can survive in a longer growing season, or that are resistant to insects or herbicides.

To date, however, most transgenic crops are largely created to resist pests and herbicides. According to the Humane Genome Project, in the year 2000, more than 109 million acres were planted with transgenic crops (almost twice the land mass of the United Kingdom), and the principal crops were herbicide-resistant and pesticide-resistant soybeans, corn, cotton, and canola (the canola plant produces canola oil).

BioFact

The first genetically modified food approved by the U.S. Food and Drug Administration was the Flavr Savr tomato in 1994. It was supposed to allow tomatoes to stay fresh longer. The gene for softening was removed, and reinserted "backwards" as a way to cancel the effect of the gene. Although the tomato may have stayed harder longer, it's not clear that it tasted any better than previous tomatoes. In fact, it might have tasted worse, because it was on the market only a very short time before it was removed.

Transgenic crops have become big business worldwide. In fact, in the United States, for certain crops they are becoming ubiquitous. The great odds are that you've eaten foods that have been genetically modified in some way today or in the last few days.

To get a better understanding of how transgenic crops are being used worldwide and how ubiquitous they have become, consider these facts:

- According to the Pew Initiative on Food and Biotechnology, 16 percent of the 670 million acres of land under cultivation worldwide in 2000 were planted with genetically modified (GM) crops.
- The amount of acreage devoted to GM crops jumped more than 20-fold in the United States between 1996–2000, from 3.7 million acres to 74.9 million acres, according to the International Service for the Acquisition of Agribiotech Applications.
- The United States is by far the biggest producer of GM crops worldwide, accounting for 68 percent of them in 2000. They were primarily soybeans, corn, and cotton, according to the International Service for the Acquisition of Agribiotech Applications. Argentina was second with 23 percent (primarily soybeans); Canada was third with 7 percent (primarily canola); and the China was fourth, with 1 percent (primarily cotton). A variety of other countries combined for the remaining 1 percent.

BioFact

It may seem odd that among the most popular transgenic crops are those that are resistant to herbicides. Why bother to bioengineer a crop that won't be killed by plant poison? Because then farmers can keep weeds under control—they can spray fields with herbicides and only the weeds will die, while the crop itself will survive. This is one of the reasons that some environmentalists oppose genetically modified crops, because they can lead to a greater use of herbicides.

- Sixty-eight percent of all soybeans grown in the United States in 2001 were genetically modified, up from 54 percent in 2000, according to the U. S. Department of Agriculture (USDA).
- Sixty-nine percent of all cotton grown in the United States in 2001 was genetically modified, according to the USDA.
- Twenty-six percent of all corn grown in the United States in 2001 was genetically modified, according to the USDA.
- In Kansas and South Dakota, 80 percent of soybean farmers grow GM soybeans. In Indiana, 78 percent grow GM soybeans, and in Nebraska 76 percent grow GM soybeans, all according to the USDA.

Using Nature's Insecticide

A particularly popular GM corn crop is called Bt corn and it uses one of nature's insecticides to protect the corn against pests.

The crop is based on the bacterium *Bacillus thuringiensis* (Bt) which produces a protein that kills caterpillars. Beginning around 1980, before the days of genetic engineering, farmers began spraying the bacteria on their crops. When caterpillars eat plant leaves that have the bacteria on them, the bacteria grow in their digestive track and produce the crystal-like Bt protein. The protein ultimately kills the insect, but it is harmless to humans and other mammals because they produce an acid that degrades the Bt protein. Because of this, Bt is a highly effective and safe natural insecticide.

Scientists isolated the gene from the bacteria that produces Bt, and they then implanted it in a variety of crops, notably corn. One of the primary insect pests that eat corn is the European corn borer (it lives in the United States, not just in Europe). When corn borers eat the corn, they die, but when humans or other mammals eat the corn, they feel just fine. For this reason Bt corn is widely used. The gene has been implanted into a variety of other crops as well. This process means that the farmers no longer need to spray their fields with bacteria to protect their crops.

BioFact

Bt corn became extremely controversial several years ago, when it was believed that it was leading to the death of monarch butterflies. Butterflies, of course, live part of their lives as caterpillars, and so an initial study claimed that Bt corn was helping wipe out large numbers of monarchs. At first, it appeared that monarchs were dying after eating milkweed plants that had accidentally been dusted with pollen from Bt corn. The study garnered a great deal of publicity, but follow-up research found that, in fact, very few monarch were dying as a result of exposure to Bt corn.

Bt does not kill every kind of insect, only caterpillars and related insects, so cannot protect against every kind of insect pest.

Modern Farms: The New Industrial Plant

The first wave of bioengineering created plants altered to withstand disease and expand the growing season—in other words, it concerned itself with creating plants for eating.

The next wave, however, will go well beyond that, and may even change our entire conception of farming—it will use plants as a lower-cost way to manufacture chemicals, industrial products, and pharmaceuticals. In fact, it's already happening. Think of it as plants as the new industrial plant—literally.

The techniques for doing this are the same as the techniques already developed for transgenic crops destined as food. A gene is isolated that produces a particular protein or family of proteins. The gene is then inserted into a crop plant. The resulting transgenic plant then produces the protein, the crop is harvested, and that protein—or else a product produced by the protein when it is used in the plant—is extracted and used.

This type of industrial farming isn't yet in widespread use and is primarily in the test phase. But it could well change the way that we produce a wide variety of products, the rural landscape, and it could affect the economics of farming.

Here are some examples of products that industrial crops have been producing. These are primarily in the test phases, and are not yet in widespread production. It is expected that they, or products like them, will eventually be mass-produced.

- **Industrial Biochemicals** Many types of industrial products used to manufacture plastics, paper, and laundry detergents are naturally occurring enzymes that aid the chemical reactions required in the manufacturing process. Because genes create enzymes and proteins, many of these industrial enzymes can theoretically be produced by transgenic plants. For example, trypsin, an enzyme used in quantities in the detergent and leather industries, has been produced by transgenic corn, as has laccase, an enzyme used to manufacture detergents and fiberboard.

BioSource

One source for learning more about the use of crops for creating industrial chemicals and pharmaceuticals is a free publication from the Union of Concerned Scientists, titled "Pharm and Industrial Crops: The Next Wave of Agricultural Biotechnology." Get it at www.ucsusa.org/food_ and_environment/biotechnology/page.cfm?pageID=1033.

- **Research biochemicals.** Laboratories and scientists require the use of biochemicals in their work, and many of these are proteins and enzymes that may be able to be mass-produced by transgenic crops. For example, the natural protein avidin, which is used to help purify other proteins, has been created by industrial crops since 1997.
- **Pharmaceuticals.** A wide variety of pharmaceuticals including hormones and drugs are being produced by industrial crops. These are then used to treat diseases including cirrhosis, cystic fibrosis, and hepatitis B and C.
- **Biologics.** Biologics are complex biological products, such as antibodies, blood products, and vaccines. These products are being produced in crops. For example, vaccines against rabies, HIV, malaria, cholera, and other diseases have been produced by transgenic tobacco, corn, and potato plants. The vaccines are purified from the plants before they are used. However, in some cases edible vaccines are being used to treat animals. Trials have been conducted in which transgenic corn produced a vaccine against transmissible gastroenteritis virus in pigs. The pigs were fed the corn and the vaccine was delivered that way.

Why We Need Biofoods

Biofoods have become one of the most contentious issues having to do with genetics, in large part because of the extremely wide gulf between its backers and detractors. Both claim the mantle of idealism and often attribute less-than-worthy motives to those who oppose them.

The backers of biofoods cite a variety of reasons why we need GM crops, including the following:

- **They can help cure world hunger.** Biofoods can be engineered to have longer growing seasons, thrive in less-than-ideal conditions, resist disease, and perhaps be more nutritious than non-GM foods. They are also cheaper to grow, allowing poor people better access to food.

BioSource

The federal government has been funding a study of one of the more novel uses of GM crops. The Department of Defense is funding a $2.3 million project at the University of Florida that will try and develop GM plants and bacteria that can detect unexploded ordnance. Unexploded ordnance leaks chemicals into the surrounding soil and the idea is that GM plants and bacteria would change color in the presence of these chemicals.

- **They can help the environment by cutting down on the amount of pesticides used.** Crops, such as Bt corn, can be engineered to produce natural pesticides that have no deleterious effect on the environment. This will help the environment because it means that fewer pesticides will need to be used.
- **They can cut down on the amount of fertilizer needed to grow crops.** Many crops, such as corn, rice, and most other plants, require a good amount of fertilizer. Runoff from fertilizers is a major source of water pollution, and fertilizer is expensive and its production uses a great deal of energy and produces greenhouse gases. Bacteria and GM crops may eventually be engineered so that the crops can use bacteria to take nitrogen from the air and make it available to the crops. Plants called legumes, which include beans, peanuts, and peas, can already do this, but most other crops can't, and therefore are treated with nitrogen fertilizers.
- **They can produce lower-cost and new drugs and vaccines.** Crops can be bioengineered to produce drugs and enzymes, potentially at a lower cost than using any other method. They could also conceivably manufacture drugs that can't be produced reliably by any other method.
- **They can help farms survive.** Every year, farmland is plowed under and turned into suburban developments, shopping malls, or office buildings because farmers are economically pressed. Bioengineering can help them produce more food at lower cost, and can therefore help farms thrive.
- **They can produce industrial products at lower cost and with less harm to the environment than manufacturing facilities.** When crops are used to produce industrial and other chemicals, they need not produce the same kinds of toxic waste and other pollution that manufacturing facilities do. Because the crops themselves manufacture the chemicals, there is no need to use other chemicals in the manufacturing process. The waste will be organic crop matter, which can easily be disposed of or recycled.
- **They may be able to produce bioplastics.** Research has been done on having GM plants create biodegradable plastics. The bacteria *Alcaligenes eutropus* can produce a kind of biodegradable plastic, and the gene responsible for it has is being transferred to corn, in an experiment to see whether corn can produce biodegradable plastic.

BioSource

If you're looking for an unbiased source of information about the issues around biotechnology and genetically modified food, there's probably no better place to turn that the Pew Initiative on Food and Biotechnology at www.pewagbiotech.org. It includes news, comprehensive reports, fact sheets, backgrounder, charts, and other information, free of hype and biased point of view.

Why BioFoods Must Be Stopped

Environmentalists and others who are against biofoods have been protesting their use for years. Although they offer a wide variety of reasons against the foods, here are some of their main concerns:

- **BioFoods may lead to an increase in antibiotic-resistant bacteria.** When creating transgenic plants, "marker" genes are used, and often those marker genes have been antibiotic-resistance genes. Some worry that the widespread use of these genes may lead to certain bacteria becoming antibiotic-resistant.
- **BioFoods may cause dangerous allergic reactions in people.** When a transgenic food is created, a new gene in it creates a protein that the food normally does not have. This can lead to unintended allergic reactions. What if the gene for creating a peanut protein were implanted into a different plant? People who were allergic to that peanut protein would not expect to find it in another food and would unknowingly eat the food.

BioFact

European countries have been particularly critical of the use of GM crops and foods. In 1998, they imposed a moratorium on the importing of any new GM foods, although existing ones were allowed to be imported. In an attempt to halt a showdown between the United States and Europe over the issue, the parliament of the European Union passed a law allowing the sale of all GM foods in Europe, but requiring that any product that contains GM foods be clearly labeled. The United States complained that the law will hurt its farmers as it would be too expensive to label the foods and because European consumers do not favor GM foods, the labels would represent an unfair trade barrier. The parliament kept the law intact.

- **Engineered plant genes may get loose in the wild.** Pollen from transgenic plants may mix with related plants, and then those genes will enter the general gene pool of plants with unknown and unintended consequences.
- **It will lead to an increase of herbicide use.** Because herbicide-resistant crops are one of the most common transgenic plants, herbicide use will increase as farmers spray the crops with herbicides that will not harm the crop, but will kill other plants and weeds.
- **Wildlife may be killed.** If plants are engineered to create bioplastics, for example, wildlife that eat those plants may be killed.

- **New plant viruses might be created.** Some plants are engineered to be resistant to viruses by implanting viral genes into their genomes. It's possible that these viral genes may combine with genes of related naturally occurring viruses and create entirely new viruses that may be more dangerous to plants than existing ones.
- **The food supply will be contaminated with transgenic crops meant for animals.** Many of the transgenic crops grown are not meant for human consumption and have not been tested to see whether they are dangerous for humans. Instead, they are tested to see whether they are dangerous for animal feed. Critics warn that there is no way to adequately keep transgenic foods out of the human food supply.

CAUTION

BioWarning

The human food supply has already been contaminated with transgenic crops meant only for animal consumption. GM corn called StarLink was only approved for animal consumption because scientists could not determine whether the corn was fit for human consumption. However, in 2000, it was discovered that the StarLink corn had ended up in taco shells and taco chips meant for human consumption. The shells and chips were recalled, but the mishap highlighted how easily GM foods meant only for animals could end up in the human food supply. Fortunately, no one was harmed.

- **It will harm farmers in the Third World.** Some GM crops are deliberately engineered so that the seeds they produce are sterile. This means that farmers will not be able to save up seeds from one year's crop to plant the next year's crop. Poor farmers in the Third World, who often live on a subsistence level, will be the ones most hurt by this.
- **It will lead to less biodiversity.** Genetic engineering of crops might lead to only a few genetic varieties of a given crop, and so will be more likely to be wiped out by disease or pests.
- **It will only help large companies.** Only large corporations have the power to create, market, and distributed transgenic plants. This could lead to an accumulation of power in their hands.
- **It may violate religious strictures.** Many religions have regulations concerning what foods can and cannot be eaten either alone or in combination. For example, Muslims and Jews both have dietary laws that ban the eating of pigs. If a gene from a pig were implanted into a vegetable, would eating that vegetable

violate dietary laws? And what about vegans, who do not eat or use any animal product? If an animal protein were produced by a plant, where does that leave them?

Not-So-Golden Rice?

As an example of how politicized the issue of genetically modified foods has become, consider the case of so-called "golden rice." The rice was developed with funding from the Rockefeller Foundation to counter a serious problem in poor Third World countries. Because many people in those countries have diets deficient in vitamin A, hundreds of thousands of children go blind every year. A variety of other conditions are caused or worsened by vitamin A deficiency in those countries, including diarrhea and respiratory diseases.

Scientists were able to insert genes from daffodils into rice, which produced beta carotene in the rice—and when beta carotene is eaten, the human body turns it into vitamin A. Beta carotene has a distinctive orange tinge, and gave the rice a golden color, hence the name "golden rice."

BioFact

The seeds of golden rice carry the gene for producing beta carotene, and so each succeeding generation of golden rice will be able to produce beta carotene. In some other transgenic plants, the transgenic plants are infertile and so cannot pass the gene to succeeding generations.

One would think that this would be one example where those who favor genetic engineering of plants and those who oppose it might agree on its benefits. Rather, it proved to be a classic instance where both proponents and opponents overstated their cases and turned a potentially helpful crop into a political football.

Before the rice had even been tested in the field to see whether it would grow properly and would deliver its benefits when eaten, the Biotech company Syngenta, which owns several patents on the rice, began making public claims about the rice's benefits. It claimed that for each month the rice's introduction was delayed, 50,000 children would go blind.

Those inflated claims drew attention to the crop and groups opposed to genetic engineering began to lobby against the rice. The environmental organization Greenpeace, for example, claimed that the rice produced too little beta carotene to be of much use and said that vitamin A deficiency could be better solved by spending money in other ways. Some in Greenpeace claimed that the development of the rice was little more than an attempt "for corporate takeover of rice production, using the public sector as a Trojan Horse."

> **BioSource**
>
> You can download a copy of the letter that the Rockefeller Foundation sent to Greenpeace about the golden rice issues at archive.greenpeace.org/~geneng/reports/food/rockletter.PDF.

The Rockefeller Foundation ultimately issued a letter to Greenpeace that in essence pointed blame at both sides for politicizing a serious public health issue. To counter Greenpeace's statements, it held that the rice could help solve the problem of vitamin A deficiency, and though, it would not be the only solution, it should still be developed. The Rockefeller Foundation also issued a public condemnation of sorts, of companies who had used the issue to their own benefit, claiming in the letter, "The public relations uses of golden rice have gone too far. The industry's advertisements and the media in general seem to forget that it is a research product that needs considerable further development before it will be available to farmers and consumers."

The controversy has not stopped research into the rice. As we write this, the rice is being tested in fields, and further research is being done on it. Scientists are cross-breeding it with local varieties of rice so that it can grow in a wide variety of growing conditions and suit a wide variety of local tastes.

A more extreme example of opposition to GM foods involves the Earth Liberation Front which in 2001 took responsibility for setting fire to a transgenic plant research building at the University of Minnesota "for the end of capitalism and the mechanization of our lives." Among the projects destroyed was one attempting to create disease-resistant crops for use by subsistence farmers in Africa.

The Least You Need to Know

- To create a transgenic plant, a gene from another plant or animal is identified and then spliced into a small ring of DNA in bacteria. The bacteria are mixed with plant cells, which takes up the gene. The resulting cells are grown into whole transgenic plants.
- In 2000, 16 percent of all crop acreage in the world was devoted to growing genetically engineered crops; 68 percent of the genetically engineered crops were grown in the United States.
- Corn, soybeans, cotton, and canola are the most popular genetically engineered crops in the world.
- Research is going on into using crops to produce industrial chemicals, vaccines, and pharmaceuticals.

- Those who favor BioFoods say that they can help cure world hunger by producing more nutritious crops that can thrive in difficult growing circumstances and be resistant to pests and disease.
- Those against BioFoods warn that they can hurt the environment, harm human and animal health, hurt poor farmers, and lead to economic concentration in the hands of a few large corporations.

Chapter 24

The Modern Pharmer

In This Chapter

- Understanding what pharming is
- How pharmers create crops and "phoods"
- A look at the economics of pharming
- What the critics say about biotech pharming corporations
- How the pharming companies defend themselves.
- The Terminator comes to a pharm near you

When we were kids, a television show used to come on every Saturday morning sometime just around dawn—*The Modern Farmer*. It showed off the latest in tractors and crop dusters, and while you may not think of it as normal children's fare, one of us (Preston) used to wake up every Saturday morning to watch. One reason was that it was the only show on TV that early in those long-ago earlier days of television. But beyond that, there was something fascinating about all that newfangled technology being used out in the vast prairies of the Midwest.

Today, there's a new kind of farmer—or more properly pharmer. This pharmer uses biotechnology to grow transgenic crops. Right now pharmers are, by necessity, big corporations and they've become controversial in recent years. In this chapter, we'll look at the economics of pharming and the controversies swirling around the big biotech companies involved in it.

What Is Pharming?

It's a picture-perfect day down on the pharm. The sun is shining in a cloudless sky after a night of cooling rain. The corn is as high as an elephant's eye—and the pharm's owner looks forward to the day in two months when he can harvest the golden crop and reap gallon after gallon of hepatitis B vaccine.

His eyes turn towards his sheep, docile as always as they graze on sweet grass. They're looking healthy as well—and every day he milks them for their yield of the protein CFTR, which can treat cystic fibrosis.

Everywhere he looks he sees transgenic plants and animals turned into nature's biotech factories, churning out vaccines, proteins, drugs, and industrial biochemicals. Sure, he longs for the day when he could go into the field on a late summer day and taste his corn kernels, wake in the morning to fresh milk from his cows, and slaughter the occasional pig for a family feast. These days, however, that's not possible because their feast would include a stew of vaccines, proteins, and biochemicals.

Progress has its costs as well as its benefits, he thinks, and he turns back to the latest transgenic plant and animal catalog. He tries to decide whether next year he should order a herd of transgenic cows for human serum albumin production, or maybe a herd of goats to produce glutamic acid decarboxylase, which helps in the treatment of type 1 diabetes.

Welcome to what might be the world of the modern pharmer. Although the scene just described is a fantasy, it's not pure fantasy. Every one of the transgenic plants or animals described has been in development one way or another over the last several years. As of this writing none of them are in actual commercial production, however many believe that the day is not far off when the American farm will produce vast amounts of pharmaceutical and other biochemicals, not just food, through *pharming*.

> **BioDefinition**
>
> **Pharming** is the use of agricultural plants and animals to produce pharmaceutical products including vaccines and proteins, as well as other biochemicals such as those used to manufacture plastics.

However, the term has increasingly been used more broadly and often refers to all agricultural biotechnology, not just biotechnology that produces vaccines and biochemicals. It is also used when farmers grow genetically engineered foods, such as Bt corn that produces its own natural pesticide.

In this chapter, we'll cover the broadest use of that term. We'll consider the pharming industry as that part of the biotech industry that applies biotech techniques to farming.

Previously in this book, we've covered genetic engineering, cloning, and the legal and ethical issues relating to the technology. In this chapter, we'll take a closer look at the economic issues involved in biotechnology in agriculture. (For information about how transgenic plants and animals are created, turn to Chapter 19.)

In the long run, the economic issues may outweigh the legal and ethical ones. Markets and economics have a way of determining the future, apart from ethical and legal issues, and it's probably economics that will ultimately determine how the pharming industry develops and how its products are used in the United States and throughout the world.

BioSource

For the main U.S. Department of Agriculture website that covers biotechnology issues, go to www.usda.gov/agencies/biotech/. You'll find information about research, laws, regulations, and much more.

A Look at Some Pharming Products

Throughout the book we've discussed a wide variety of transgenic plants and animals being used for a variety of purposes, so we won't cover them all again.

When it comes to pharmaceuticals, there are an astonishingly wide variety of products that have been tested or are being considered. To give you a small taste, look at the following table, taken from the U.S. Department of Agriculture (USDA) website. It details transgenic mammals in development, the drug or protein the animal will create, and how that protein or drug will be used. Keep in mind that this list is from back in 1999 (there is no more recent list), so there are certainly a wider variety of transgenic animals being planned today. Also, the list doesn't include mice, where the variety of transgenics is immense.

Animal Pharming Pharmaceutical Products in Development as of 1999

Animal	Drug/Protein	Use
Sheep	Alpha1 anti trypsin	Deficiency leads to emphysema
Sheep	CFTR	Treatment of cystic fibrosis
Sheep	Tissue plasminogen activator	Treatment of thrombosis

continues

Animal Pharming Pharmaceutical Products in Development as of 1999 (continued)

Animal	Drug/Protein	Use
Sheep	Factor VIII, IX	Treatment of hemophilia
Sheep	Fibrinogen	Treatment of wound healing
Pig	Tissue plasminogen activator	Treatment of thrombosis
Pig	Factor VIII, IX	Treatment of hemophilia
Goat	Human protein C	Treatment of thrombosis
Goat	Antithrombin 3	Treatment of thrombosis
Goat	Glutamic acid decarboxylase	Treatment of type 1 diabetes
Goat	Pro542	Treatment of HIV
Cow	Alpha-lactalbumin	Anti-infection
Cow	Factor VIII	Treatment of hemophilia
Cow	Fibrinogen	Wound healing
Cow	Collagen I, collagen II	Tissue repair, treatment of rheumatoid arthritis
Cow	Lactoferrin	Treatment of GI tract infection, treatment of infectious arthritis
Cow	Human serum albumin	Maintains blood volume
Chicken, Cow, Goat	Monoclonal antibodies	Other vaccine production

The Economics of Pharming

The United States has a strong agricultural heritage and some of its founding farmers, notably Thomas Jefferson, envisioned the country as one made up of small independent farmers. In fact, throughout much of the country's history, small family farms formed its backbone.

The current pharming industry, however, is anything but small by necessity. It has to spend many millions of dollars a year in research and much of that money will never lead to usable products. After it produces potentially usable products, it has to pay for expensive trials that are regulated by a variety of different government agencies.

BioFact

In the instance when a gene is used to create protein secreted into milk, the gene is present in all of the animal's cells. So why does the animal only secrete proteins in an animal's milk? The gene is attached to a piece of regulatory DNA that is designed to express genes only in mammary tissues. The gene is present everywhere but the protein is only produced in breast cells.

It also has to pay for marketing costs and ultimately the costs of producing the products. To a great extent, the business is based on gambles because there's no guarantee that the companies will come up with useful products that will be approved by the government. Even if they do, there's no way to guarantee that consumers will want to buy those products.

The Animal and Plant Health Inspection Service (APHIS) of the U.S. Department of Agriculture, which regulates bioengineered food, estimates that the cost for making a single transgenic animal ranges from $20,000 to $300,000, and "only a small portion of the attempts succeed in producing a transgenic animal," it goes on to note.

So why are so many companies involved in pharming? The benefits could be enormous. According to an estimate by the *Financial Times*, a single herd of 600 transgenic cows could meet the entire worldwide demand for certain pharmaceuticals. An example of this is human serum albumin, which is used to treat burns and traumatic injuries.

Even though startup costs for pharming are substantial, in the long run it holds out the promise of creating pharmaceutical products at far less cost than creating them by traditional means. Richard McCloskey, vice president of medical research for the biotechnology firm Centocor, said at a forum sponsored by the Pew Initiative on Food and Biotechnology that it costs $80 million to manufacture 300 kilograms of antibodies in the current traditional manner, but that transgenic corn could produce the same amount of antibodies for far less money—$10 million.

BioFact

The pharming industry is still young and a lot of its basic economics are still uncertain. For example, no one is still sure how much land will need to be devoted to farmland for transgenic crops. However, there are some estimates. Centocor's McCloskey believes that 15 kilograms of plant-produced antibodies can meet the needs of 10,000 patients in a year. It would take an acre and a half of transgenic corn to produce that amount of antibodies in a year, he estimates.

By McCloskey's estimates, corn may be the most cost-effective transgenic animal or plant for producing antibodies. To produce 300 kilograms of antibodies from tobacco, he estimates, would cost approximately $20 million, and to produce that same amount from goats would cost nearly $60 million.

Is Bigger Better for Poor Farmers?

Some people claim that the very size of the biotechnology companies means that they are bad for agriculture. Such claims are not generally directed at production of pharmaceuticals but rather at transgenic food crops. These people contend that bigger is not better—it's worse. And they say that the biotechnology industry has been using its financial clout to harm agriculture, particularly in the poor countries of the world.

The advocacy group ActionAid claims in a study titled "GM Crops—Going Against the Grain," that genetically modified crops has have harmed poor farmers in the Third World, and that "The expansion of GM (genetically modified) crops is more likely to benefit rich corporations than poor people."

The report claims that the practices of bioengineering companies can push poor farmers deeper into poverty. For example, it notes that the chemical giant Monsanto sells a herbicide called Roundup, and also sells "Roundup Ready" GM seeds that grow plants that are resistant to Roundup. The idea is that farmers plant the Roundup Ready crops that are resistant to the Roundup herbicide, then spray the fields with Roundup, which kills weeds but not the crops. However, poor farmers have a difficult time affording the expensive herbicides and would be better off farming non-GM crops, says the report.

BioFact

The ActionAid report makes many claims against biotech giants. These claims can't be verified, but here are some of what the report claims: Only one percent of GM research is aimed at crops used by poor farmers in poor countries; only a very small range of GM crops that might be used by poorer farmers are being researched, "but they stand only a one in 250 chance of making it into farmers' fields;" and that 91 percent of all seeds for GM crops grown worldwide in 2001 were from a single company, Monsanto.

According to the study, "GM crops are unlikely to help eradicate poverty because yields seem to be no more than non-GM crops and sometimes need more chemicals."

It goes on to say:

> "Insecticide use on GM cotton has fallen in some locations, but these gains may be short-lived as insects develop resistance to the insecticide that the cotton expresses. In time, farmers may need to invest in more, not fewer, chemicals. This also applies to chemical use on herbicide-resistant GM crops, which has gone up rather than down as farmers use chemicals more frequently and/or in greater amounts. Herbicide use per hectare in Argentina has more than doubled on GM fields compared to conventional varieties."

It concludes, "GM could be disastrous for small-scale farmers as the costs are much higher and they risk falling into debt." On the other hand many farmers, both large and small scale, have voted for GM crops with their pocketbooks. Among the advantages are improved disease and insect resistance that may not show up in test plots.

A Seedy Story?

One biotech company practice that many critics bemoan is the practice of not allowing farmers to save seeds from one harvest to the next that can be used in subsequent seasons. In poor countries in particular, farmers take seeds from one year's crop and plant that seed the following season. This allows them to save a substantial amount of money because instead of having to buy all new seeds from one year to the next, they can instead use seeds from the previous year's harvest.

However, pharming firms typically do not allow seeds to be saved. They require that new seeds be bought each season. According to ActionAid:

> "Before (farmers) can obtain and use the seeds, farmers have to sign a contract with the company obliging them to pay a royalty or technology fee, to agree not to save or replant seeds from the harvest, to use only company chemicals on them and to give the corporation access to their property to verify compliance."

The result, according to ActionAid, is that:

> "Having to buy external supplies of seeds and pesticides leaves farmers more economically and agriculturally dependent on corporations. The technology fee makes such seeds prohibitive for the poorest farmers who lack access to credit. The contracts are complex and easily misunderstood by farmers, especially those who are illiterate."

BioSource

For a copy of the ActionAid report, go to www.actionaid.org/resources/pdfs/gatg.pdf.

The Pharmers Disagree

The pharming industry disagrees with ActionAid and other critics. They claim that their products, such as insect- and disease-resistant crops, help poor farmers by increasing yields under real-world conditions where crops are often ravaged by insects and disease. Some GM crops can also be grown with fewer expensive pesticides and herbicides. And that, they say, ultimately means lower food costs and the ability to feed the hungry of the world more easily. They claim that pharming may have as much of an impact on food production as did the previously heralded "Green Revolution," which increased crop yields throughout the world. And nobody is forcing farmers to buy GM seeds—if they agree with GM critics they can just continue farming rather than begin pharming.

The International Food Policy Research Institute has done a study which contradicted the ActionAid study—it shows that transgenic crops can be a boon to poor Third World farmers. It claims that transgenic crops "could increase the productivity of several crops, including the two major staples, rice and wheat, by 15 to 20 percent," and adds, "A World Bank panel has estimated that transgenic technology can increase rice production in Asia by 10 to 25 percent in the next decade."

The report concludes:

"The next generation of crops with improved output traits could confer nutritional benefits to millions who suffer from malnutrition and deficiency disorders … The Nuffeld Council on Bioethics recently concluded that a compelling moral imperative exists to make transgenic crops available to developing countries that want them to combat hunger and poverty."

BioSource

For more information about the International Food Policy Institute, go to www.ifpri.org/.

The Terminator Stalks the Farm

You thought it was frightening having the Terminator run for Governor of California? Get ready for something even more frightening—the Terminator comes to a farm near you.

No, it's not a pumped-up Arnold Schwarzenegger we're talking about. It's a controversial technology called the "Terminator" that produces sterile crops so that their seeds cannot be replanted.

Monsanto is the most well-known of the companies that developed the technology. As discussed earlier in this chapter, companies like Monsanto do not want farmers to be able to use seeds from their genetically modified crops in future years. They want farmers to have to buy seeds from them every year.

Even though the companies frequently force farmers to sign contracts saying that they will not plant the seeds in future years, there is the chance that farmers will not obey the contracts. So the biotech firms came up with the idea for Terminator technology. The idea is simple: The crops will produce only sterile seeds so that even if a farmer wanted to replant them, it wouldn't help the farmer. Nothing would grow.

Several years ago the technology caused a major uproar and not just among consumer and environmental groups. The World Bank and the Rockefeller Foundation both opposed the selling of Terminator seeds. The World Bank warned that it could lead to economic disaster for millions of farmers and widespread famine.

"The small farmers in the developing world who still rely extensively on their ability to hold back their seeds … who can get wiped out for one bad season, would suddenly find themselves with no seeds for the next year and no money to buy new seeds," warned Ismael Serageldin of the World Bank.

Another worry about the technology is that if the Terminator crop cross-pollinated with a normal crop, it could render the normal crop sterile.

Bowing to public pressure, Monsanto agreed not to sell the seeds. However, it never completely ruled out not using the technology. There are those who worry that Terminator the Seed 2 may ultimately occur. A group called the International Seed Association, made up of large agricultural businesses, has come out in favor of the technology, so it may live yet again.

BioFact

The biotech companies didn't invent the Terminator technology by themselves. Also involved in the research is the U.S. Department of Agriculture (USDA). In fact, the USDA is a patent holder for some Terminator technology.

The Least You Need to Know

- In the narrowest sense, pharming refers to the use of transgenic agricultural plants and animals to create drugs and other pharmaceutical products—although in a broader sense it refers to the use of agricultural transgenic technologies.
- There have been many tests or planned tests of transgenic goats, pigs, cows, corn, and other plants and animals to produce pharmaceutical products.

- Those in the biotech industry believe that pharming can produce pharmaceuticals and drugs at much lower cost than can be produced by traditional means.
- Pharming critics contend that agricultural biotech companies might harm poor farmers in Third World countries by selling them expensive GM seeds and agricultural chemicals that they cannot afford and do not need.
- Biotech companies claim that their technology can feed the hungry by producing more nutritious crops at lower cost.

Resources for More Information

Books and the web are the best places to get more information about cloning, DNA, and genetics. Here are the best places to turn.

Genetic Diseases

GeneMedNetwork
genemed.org/
Site that lets you search genetics resources for information about genetic therapy, diseases and research.

Yahoo! Listing of Sites Related to Genetic Disorders
dir.yahoo.com/Health/Diseases_and_Conditions/Genetic_Disorders/
Yahoo! listing of websites devoted to numerous genetic disorders and their cures.

National Organization for Rare Disorders
www.rarediseases.org
Website of the National Organization for Rare Disorders, which has a great deal of information about genetic disorders.

National Society of Genetic Counselors
www.nsgc.org/
Website of the National Society of Genetic Counselors, who offer genetic advice to individuals.

Genetic Alliance
www.geneticalliance.org/
Educational alliance funded by a broad range of governmental organizations, non-profit institutions, and biotech firms that provides information about genetics.

GeneTests
www.genetests.org/
Publicly funded site that offers information about genetic testing to physicians, healthcare providers, and researchers.

Cloning

New Scientist Cloning Section
www.newscientist.com/hottopics/cloning/cloning.jsp
Articles from the *New Scientist* magazine about cloning.

HumanCloning.org
www.humancloning.org/
Organization devoted to promoting human cloning.

Usenet Cloning Newsgroup
alt.bio.technology.cloning
UseNet discussion group focusing on cloning and related issues.

Americans to Ban Cloning
www.cloninginformation.org/
Website of Americans to Ban Cloning, a pressure group that wants to ban human cloning and therapeutic cloning.

Clonaid
www.clonaid.com
Online site of the company run by the Raelian UFO cult which claims to have cloned human beings.

Clone Rights United Front
www.clonerights.com
Group devoted to legalizing and promoting human cloning.

Yahoo! Listing of Cloning Articles
d1.dir.scd.yahoo.com/science/biology/genetics/cloning/human_cloning/
Yahoo! listing of current cloning articles.

National Academy of Sciences Report on Cloning
www7.nationalacademies.org/cosepup/Human_Cloning.html
Report about the scientific and medical aspects of cloning, by the National Academy of Science.

Bioengineering and Agriculture

Animal Biotechnology
www.animalbiotechnology.org/
Information and news about the use of biotechnology on animals in agriculture.

AgBiotechNet
www.agbiotechnet.com/
Site that promotes the use of biotechnology in agriculture, with many articles about biotech in agriculture.

The Alliance for Better Foods
www.betterfoods.org/
Organization devoted to promoting the use of genetically modified foods.

Center for Ethics and Toxics
www.cetos.org/
Website of the Center for Ethics and Toxics, an environmental group concerned with the possible dangerous effects of biotechnology as well as pesticides and chemical toxins.

Council for Biotechnology Information
www.whybiotech.com/
Website run by a the Council for Biotechnology Information, which promotes the use of agriculture biotechnology and genetic modification of foods.

Alliance for Bio-Integrity
www.bio-integrity.org/
Site of the Alliance for Bio-Integrity, which is opposed to the use of genetically modified foods.

Organic Consumers Association
www.organicconsumers.org/
Website run by the Organic Consumers Association, which promotes the use of organic foods, and opposes the use of genetically modified foods.

Information and Organizations About Cloning, DNA, and Bioengineering

Center for Genetics and Society
www.genetics-and-society.org
Nonprofit institution devoted to encouraging responsible uses and governance of human genetic and reproductive technologies.

HowStuffWorks
www.howstuffworks.com
Many articles explaining how cloning, DNA, and cells work.

actionbioscience.org
www.actionbioscience.org/
Non-profit, educational site devoted to promoting literacy in the biosciences.

BBC Articles about Genes, DNA and Cloning
www.bbc.co.uk/science/genes
Collection of articles from the British Broadcasting Company about genes, DNA, and cloning.

Council for Responsible Genetics
www.gene-watch.org/
Public policy group that takes a frequently critical view of genetic and cloning technologies.

Genome Programs of the U.S. Department of Energy Office of Science
www.doegenomes.org
Probably the best website you can find anywhere for information about DNA, genetics, and cloning. It's run by the federal U.S. Department of Energy Office of Science, founder of the Human Genome Project.

Wired Magazine Biotechnology Articles
www.wired.com/news/medtech
Section of *Wired Magazine* devoted to medical technology, including biotechnology and cloning.

Dolan DNA Learning Center
www.dnalc.org/
Education and information site run by the Dolan DNA Learning Center of the well-known biotech research laboratory Cold Spring Harbor.

Genetic Science Learning Center of the University of Utah
gslc.genetics.utah.edu/
Online site of the Genetic Science Learning Center of the University of Utah—excellent educational resource for parents and students.

GenomeWeb
www.genomeweb.com/
News about genetics-related technology.

National Public Radio's DNA Files
www.dnafiles.org/
Series of programs from National Public Radio about DNA, by reporter John Hockenberry.

GeneLetter
www.geneletter.com/
Online magazine covering genetic medicine, society and culture, run by the private company GeneSage.

GeneSage
www.genesage.com/
Private company that provides information services about genetics to doctors, health-care providers, and consumers.

PBS/Scientific American Gene Hunters TV series
www.pbs.org/saf/1202/
Online site of the PBS/Scientific American Gene Hunters TV series.

DNA Interactive
www.dnai.org/
The DNA Interactive website, an educational site run by well-known biotech research laboratory Cold Spring Harbor.

Biotechnology Industry Organization
www.bio.org/
Website of the Biotechnology Industry Organization, an industry biotechnology trade organization.

Books

Human Cloning and Human Dignity: The Report of the President's Council on Bioethics. Public Affairs Reports, 2002.

Borem, Aluizio, Fabricio R. Santos, and David E. Bowen. *Understanding Biotechnology.* Prentice Hall, 2003.

Calladine, C.R. and Horace R. Drew. *Understanding DNA: The Molecule and How It Works*. Academic Press, 1997.

Gonick, Larry. *The Cartoon Guide to Genetics*. Perennial, 1991.

Hawley, R. Scott and Catherine A. Mori. *The Human Genome, A User's Guide*. Harcourt Academic Press, 1999.

Hoagland, Mahlon, Bert Dodson, and Judith Hauck. *Exploring the Way Life Works: The Science of Biology*. Jones and Bartlett, 2001.

Kristol, William and Eric Cohen, eds. *The Future is Now: America Confronts the New Genetics*. Rowman & Littlefield, 2002.

Peters, Ted. *Playing God? Genetic Determinism and Human Freedom*. Routledge, 1997.

Rantala, M.L. and Arthur J. Milgram, Ph.D., eds. *Cloning For and Against*. Open Court, 1998.

Ridley, Matt. *Genome*. HarperCollins, 2000.

Tagliaferro, Linda and Mark V. Bloom, Ph.D. *The Complete Idiot's Guide to Decoding Your Genes*. Alpha Books, 1999.

Tudge, Colin. *The Impact of the Gene: From Mendel's Peas to Designer Babies*. Hill and Wang, 2002.

Appendix B

Glossary

adenine One of the four nucleotides that make up DNA. Typically it is written as A.

allele One of the forms of a gene for a given trait.

amino acids The building blocks of proteins. The human body can manufacture many amino acids, but not all of them in large enough quantities, and so these have to be provided in our diet. Twenty different kinds of amino acids make up proteins, by combining together in different combinations.

amniocentesis A procedure in which fluid is taken from the amniotic sac which surrounds an unborn child, and is then analyzed. The fetus sheds cells into the fluid and these are tested for genetic disease.

antibody A protein in humans and other mammals that protects against disease by fighting infections and foreign bodies.

antigen A foreign substance, often part of invading bacteria or viruses, that stimulates an immune response within your body.

antisense therapy A kind of gene therapy in which a faulty gene is stopped from producing a defective protein.

autoimmune disease A type of disease in which the body's immune system mistakenly attacks the body that it was designed to protect. Rheumatoid arthritis and multiple sclerosis are two examples of autoimmune diseases.

base A portion of the DNA molecule. DNA has four bases: adenine (A), thymine (T), cytosine (C), and guanine (G).

base pair Two bases that combine and form a "rung" on the double helix DNA molecule. Adenine (A)can only bond with thymine (T), and cytosine (C) can only bond with and guanine (G).

behavioral genetics The study of the relationship between genes and behavior.

biochip A microchip that can be used to diagnose genetic diseases. Also called DNA chips, DNA arrays, and microarrays.

bioengineering Manipulating the genes of plants or animals for a specific purpose, for example to create crops resistant to frost.

bioethicist Someone who studies ethical issues in medicine and genetics, such as issues around cloning and bioengineering.

biofoods Foods that have been changed by genetic engineering.

carcinogens Substances that have the potential to cause cancer.

carrier A person who has a recessive gene for a genetic disease, but does not develop the disease because he or she also has a dominant partner gene that overrules the recessive gene.

cell division The dividing of a cell to produce two new cells. In cell division, the DNA is duplicated so that identical DNA will be present in each new cell.

cell fusion When two cells combine to become one cell. A sperm fertilizing an egg is an example of cell fusion.

cell line Cells grown outside living organisms from which they were taken. These cells are duplicated from generation to generation, and so are called a line.

cell membrane The outside covering of a cell that contains it and protects it, but allows nutrients and other materials to pass through.

centromere The body in the center of a chromosome that helps move it into the new cells created during cell division.

chimera A cloned piece of DNA with pieces from more than one source—for example, from a virus and a plant.

chloroplasts Small organs in plant cells that produce chlorophyll, which converts sunlight into energy and produces oxygen.

chorionic villus sampling (CVS) A procedure that screens a fetus for disease. In CVS, cells are taken from the placenta and examined for genetic disease.

chromatin Material that makes up chromosomes, composed of approximately one half DNA and one half protein.

chromosomes Thread-like structures in a cell's nucleus that contains its DNA.

clone Organisms that have the exact same DNA genomes. Human identical twins are clones, as are animals that have been created using another organism's DNA, such as the sheep Dolly.

cloning vector A virus or other kind of material that can be used to insert foreign DNA into a cell.

codon A group of three consecutive nucleotides that taken together represents a specific amino acid for when proteins are built from DNA.

crossing over The process of genes being swapped on chromosomes during meiosis when egg and sperm cells are produced.

cytoplasm The liquid part of a cell in which all small bodies such as the nucleus are suspended.

cytosine One of the four nucleotides that make up DNA. Typically it is written as C.

DNA (deoxyribonucleic acid) The basic molecule of heredity that contains all genetic material. It forms the basis of life and is passed down from generation to generation.

DNA fingerprinting Also called DNA profiling or DNA typing, a process in which a given DNA sample is compared to another to help establish identity, for example to establish paternity or whether someone may have committed a crime.

DNA sequencing The process of determining the exact order of the base pairs that make up the DNA in a given organism.

dominant A form of a gene that can override its partner gene. For example, in eye color the gene for brown eyes is dominant over blue eyes so a person with one gene for brown eyes and one for blue will have brown eyes.

double helix The shape of the DNA molecule, made up of two spirals, with steps connecting them, something like a spiral staircase.

embryo twinning A method of cloning animals in which a developing embryo is split into two at an early enough stage so that each embryo will develop into an identical animal.

endoplasmic reticulum The part of a cell in which proteins are manufactured.

enzyme A large molecule that speeds up a chemical reaction.

eukaryote A cell with a nucleus.

exon The part of a gene that contains the string of DNA bases that is decoded to create a protein.

gene The basic unit of heredity. A gene is a discrete portion of DNA. A single gene creates a single protein or family of proteins.

gene therapy A form of medicine in which genes are manipulated in order to correct diseases caused by genetic defects.

genetics The branch of science concerned with the study of heredity.

genome The sum of all the DNA on the chromosomes in a given organism. Every individual plant and animal has its own distinctive genome.

genomics The study of genomes.

genotype The actual genetic makeup of an organism—the genes that an organism has. Two individuals that have the same external trait may have different genotypes. For example, two people may both have brown eyes, but one of them may have one gene for brown eyes and one gene for blue eyes, and the other may have two genes for brown eyes. In that case, they would have different genotypes. *See* phenotype.

germ cells The reproductive cells in an organism—the sperm and egg cells.

Golgi apparatus The part of a cell that does the final "packaging" of proteins before they are ready to be used or transported.

guanine One of the four nucleotides that make up DNA. Typically it is written as G.

helix A spiral. The DNA molecule is a double helix made up of two spirals with steps in between, like a spiral staircase.

heredity Passing of traits from one generation to the next.

heterozygous When the two alleles for a given gene in an animal or plant are different.

homozygous When the two partner alleles of a single gene in an animal or plant are the same.

hormones Chemicals produced by the body that affect the growth and function of cells, organs and the body.

Human Genome Project A joint scientific effort in which the entire human genome was mapped.

hybrid An organism that is formed when two different species are combined.

in vitro fertilization (IVF) A means of reproduction in which sperm and eggs are mixed together outside the body, and the resulting embryos are implanted in a mother.

insulin A protein that regulates the level of sugar in the blood. People with diabetes do not produce enough insulin on their own, and so need to get injections of insulin, or get it in another way. Scientists have bioengineered animals to produce human insulin in their milk, as a way of mass-producing the protein.

interferons Proteins that help the body fight off disease by aiding the body's natural defenses.

intron A internal portion of a gene that does not contain the codons that direct protein synthesis.

lysosomes Tiny enzyme-filled sacs in a cell that break down and digest things no longer needed in the cell, or that are cell invaders.

meiosis A process in which mature sperm and egg cells are formed.

Messenger RNA (mRNA) A kind of RNA that carries the blueprint for the protein from the DNA in the nucleus out to the ribosomes.

mitochondria Organelles that produce energy for cells. Because they contain their own DNA and a covering membrane, some people believe they may have originally been bacteria that formed a symbiotic relationship with larger cells, and ultimately became part of them.

mitosis A process in which the cell's chromosomes are duplicated during cell division.

monogenic disease A disease caused by a mutation of a single gene.

mutation A change in the normal structure of a gene. Mutations lead to proteins or families of proteins being created incorrectly or not at all.

nuclear membrane A protective covering around the nucleus of a cell, protecting its chromosomes, and allowing certain materials to pass in and out of the nucleus.

nuclear transfer A process of cloning in which the DNA from an adult cell is put into an egg cell that has had its nucleus removed. The resulting embryo is then implanted in an animal and brought to term. This is how Dolly the sheep was cloned.

nucleotide A building block of DNA. There are four nucleotides: adenine (A), thymine (T), cytosine (C), and guanine (G). Adenine (A) can only bond with thymine (T), and cytosine (C) can only bond with and guanine (G).

nucleus The central part of the cell that holds its chromosomes.

oncogene A gene, usually mutated or expressed abnormally, that can cause cancer.

phage A virus that attacks the bodies of bacteria. Phages can also be used to inject genes into cells.

pharming A somewhat slang term used to describe the process of bioengineering plants and animals so that they produce human proteins that can be used to treat disease.

phenotype The physical manifestation of an organism's traits. Two different organisms can have the same phenotype but a different genetic makeup. For example, two people may both have brown eyes, but one of them may have one gene for brown eyes and one gene for blue eyes, and the other may have two genes for brown eyes. In that case, even though their genetic makeup is different, they would be of the same phenotype for that trait.

point mutation A mutation that occurs when a single nucleotide is changed permanently within a gene.

polygenic disease A disease caused by a mutation of more than one gene.

prokaryote A one-celled creature lacking a nucleus, such as bacteria.

promoter A segment of DNA before a gene that turns the gene on or off to produce a protein or family of proteins.

protein synthesis The making of proteins by assembling amino acids in an ordered chain.

recessive A gene whose function in inactive if there is also a dominant gene present. It functions only if two recessive genes are present.

recessive disease A disease that only occurs if two copies of a recessive gene are present in the organism.

Retrovirus A kind of virus that has RNA instead of DNA for its genetic material.

Ribonucleic acid (RNA) A kind of nucleic acid, related to DNA, that carries out a number of activities related to protein-building.

Ribosomes Particles within a cell that build proteins.

selective breeding Crossing animals or plants with each other to produce a desired physical characteristic.

somatic cells All of the cells in a body except for the reproductive cells.

stem cells Embryonic cells that are still undifferentiated and can develop into any kind of cell in the body. Many researchers believe that stem cells might be used to cure many diseases because they can be used to regenerate damaged nerve tissue, for example.

telomere A structure on the end of a chromosome that protects it. Each time a chromosome divides, its telomere may shorten slightly. Because cloned animals are made from DNA that has already divided many times, the animal's telomeres are already shortened. Scientists believe this may be the reason that clones are susceptible to a variety of health problems, including premature aging.

thalassemia A genetic disease in which it is difficult or impossible for cells to produce the protein globin. Globin is a component of hemoglobin, which carries oxygen in our red blood cells, and so those who suffer from the disease have abnormally small red blood cells and carry little oxygen.

thymine One of the four nucleotides that make up DNA. Typically it is written as T.

transcription The act of transcribing the information in DNA to an RNA message strand, which in turn carries the information required to create a protein.

transgenic animal An animal that is made up of genes from more than one organism.

transgenic plant A plant that is made up of genes from more than one organism.

vacuoles A part of a cell that stores and processes nutrients and wastes.

vector In gene therapy, a way of delivering a gene into an existing organism's cells.

xenotransplantation Using animal organs, such as organs from pigs, to replace human organs that are damaged.

Index

A

B

C

D

E

F

G

H

I-J

K-L

M

N

O

P–Q

R

S

T

U-V

W

X-Y-Z